AMAZON ORIGINAL

The Grand Tour

A-Z
OF THE
CAR

HarperCollins*Publishers*

COME

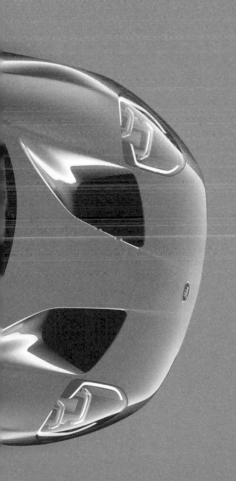

Welcome to the only compendium of car facts you'll ever need, assuming you don't need too many car facts. During the making of *The Grand Tour* television programme we're forever being asked the same questions. 'How was Daihatsu established?' 'What killed the founder of MG?' 'Is there a name for the massive junction near Birmingham where the M40 and M42 motorways meet?' And, as always, we say, 'Shut up, fictional question-asking person we've just invented for rhetorical purposes. It'll all be in the book of car facts we're going to compile.' And now, at last, the answers to these and many, many other motoring-related questions have been lovingly assembled into what we can only call an actual encyclopaedia of the car, although one of those encyclopaedias where there might be some things that the authors have forgotten and bits of it are made up and James got bogged down in certain facts even though we asked him not to. If this sounds like the kind of encyclopaedia you need, then welcome, friend, welcome. Otherwise, it's too late. You were given this as a present and to return it would look rude. So there.

A

ALFA ROMEO

AXLE TRAMP

ARIEL

ASTON MARTIN

AMAZED

AUDI

AUSTIN

AMC

ABS

AUTOTESTING

AUTOMATIC GEARBOX

A IS FOR ...

Ha! I am steering und you are not.

ABS

Also known as anti-lock braking system, ABS is a feature which detects if any of a car's wheels are about to lock under braking and rapidly releases then re-applies braking pressure many times per second to hold the wheel on the point of locking, and therefore maximum braking effect, while keeping the wheels turning, thereby allowing the car to steer rather than skid. Sounds pretty boring, but for some reason certain men's magazines are obsessed with it and put ways to improve your ABS on the cover every month. 'Better ABS in just three weeks' they say, or 'Get bigger ABS this summer'. Quite why you'd want bigger anti-lock brakes isn't clear, especially in summer when, actually, the risk of skidding is likely to be lower due to drier weather.

AMAZED

State of perpetual wonderment permanently experienced by British MOT testers upon being presented with an old and not very interesting car, at least according to the adverts for old and not very interesting cars. E.g. '1983 Morris Ital. This car is in exceptional condition for its age. In fact, at its last MOT the tester was amazed to see one in this condition...'

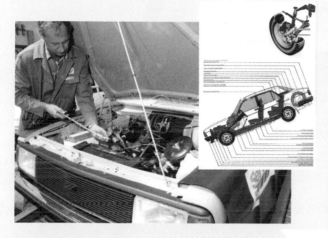

APEX

A point on the inside of a corner which should be touched if a driver is following the racing line. Technically, the plural of apex is 'apices' but this is best avoided unless you are taking the racing line towards sounding like a complete knob.

ARIEL

Delightful West Country-based maker of cars for people who think Caterhams are too luxurious. And not draughty enough. Now boasts a two-car range, with the Atom for people who want to feel the wind up their trouser legs and the Nomad for people who'd like to get covered in mud.

AMBER GAMBLER

Someone who upon seeing an orange traffic light accelerates to get through before the light turns red. A normal person, basically. Amber gambler is one of those old-fashioned scaremongering terms used in 1950s adverts, a bit like 'ne'er-do-well' and 'communist'.

ARRESTER BED

Also known as a runaway truck bed. One of those long pits full of gravel at the side of the road on a steep downhill section, designed to be used by out-of-control lorries. But let's be honest, who hasn't thought, 'That looks like fun. I might just pile my car into one to see what happens.' Well, what would happen is that it would probably break your car a bit. But hey, if you want to have a crack, knock yourself out. Oh yeah, that's something else that might happen too. Do let us know.

A IS FOR ALFA ROMEO

Possibly the most Italian of all car companies, founded in 1910 by a Frenchman. Over the years Alfa Romeo has made many legendary cars including the Spider, the Giulia, the Tipo 33 Stradale, the Montreal, the Alfasud, the GTV6, the SZ and the 156, all of which have been a heady cocktail of joyful sights and giddy sounds skilfully blended with profound disappointment. There was also the Arna, which married Japanese style to Italian engineering and was therefore precisely the wrong way round, and the bizarrely named Alfa Romeo Romeo, which was both the Boutros Boutros-Ghali of the roads and also a van. No one remembers that Alfa used to make vans. Their stock in trade, however, was sporting cars, the best of which made a parping noise from the exhaust (see rasping) and caused grown adults to get misty-eyed about 'soul' and 'character' right up until the moment they shouted, 'the bloody gearbox has broken again'.

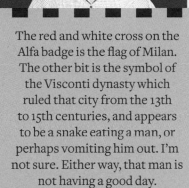

A IS FOR AMC

American Motors Corporation. US car company famed for several things, one of them being a cack-handed product-placement deal with the 1974 Bond film *The Man with the Golden Gun*, in which a load of left-hand-drive American cars were shipped to Thailand, even though they drive on the left side of the road, and then one of them did a cool spiral jump that was completely ruined by a stupid sound effect. AMC is also famous for giving its cars funny names like the Gremlin, the Matador and the Rambler Rogue. They also made a muscle car called the AMC Machine, which is a brilliant name for a muscle car when you think about it. Sadly, the company got a lot of stick for making the infamous Pacer and not so much credit for coming up with the Eagle, which was the inadvertent prototype for all modern crossover cars, and eventually suffered the terrible fate of being bought by Renault and then the blessed relief of going out of business with only their Jeep division escaping from the wreckage.

TRUE FACT: The 'flap' door handle of the Morris Marina, famously used on everything from Range Rovers to Lotus Esprits, appears to have been cribbed from a very similar design first used on AMCs in the 1960s.

A IS FOR AMG

German tuning company founded in 1967 by former Mercedes engineers Hans Werner Aufrecht (that's the A) and Erhard Melcher (that's the M), initially working out of Aufrecht's house in Großaspach (and there's the G). Began making racing cars and then moved into tuned and re-styled road cars. Developed a reputation as the leading tuning house for making Mercs go 'BRRRRRUUUUUUBBBUBUUBBUBUB'. Taken over by Daimler-Benz in 1999 and became the company's in-house department for making Mercs go 'BBBBBBUUUUUUBBBUBUUBBUBUB'.

A IS FOR ASTON MARTIN

British sports car company named after co-founder Lionel Martin and the Aston Clinton hill climb course in Buckinghamshire. It's a good job the co-founder wasn't called Peter Escape and regularly raced at Brands Hatch because that would have been confusing. Throughout its illustrious history Aston Martin has made two things: 1. Cars that are very delightful but not actually very good and 2. No money. That is until that nice Andy Palmer chap came along and led them to the DB11 and the new V8 Vantage, both of which are actually very good indeed. Also, there's some ghastly whisper that they might be turning a profit. Someone needs to have a word with this Palmer maniac before it's too late.

LLO 840D

V12 AML

Good news everyone, we got through a thing about Aston Martin without once mentioning James bloody Bon ... BAH!

A IS FOR AUDI

German car maker with a four-ring logo to mark that it's actually made up of four car companies – Audi, DKW, Horch and Wanderer – which merged in 1932. In 1969 they also sucked in Wankel radicalness enthusiasts NSU but presumably couldn't be bothered to add another ring. Over the years, Audi has been in the vanguard of many innovations including aerodynamics, galvanised bodyshells, four-wheel drive for performance cars, not using the indicators at all, and driving far too close to the car in front like a total anus.

A IS FOR AUSTIN

Defunct British car company founded by Herbert Austin in 1905 and based in an old print works in Longbridge, which was chosen because in those days it was in the countryside rather than, as now, in Birmingham, and that meant the gritty city air wouldn't spoil the finish of newly painted cars. TRUE FACT. Once such a thriving concern that Longbridge was Britain's largest factory and its production line the longest in Europe. The Austin name was respected and attached to cars like the Mini and the Seven (which was licensed around the world and helped to launch both BMW and Nissan). Unfortunately British Leyland set in, it became attached to words like 'Allegro', and from then on it was doomed. The Austin officially died in 1987 when the boss of Austin Rover decided the company sounded better as just 'Rover' and now it exists only as a trademark owned by a company in China.

A IS FOR ...

AUSTRALIA

Rugged, outdoorsy nation that developed its own car industry making rugged, outdoorsy cars. This was entirely down to necessity, because conditions in the Outback demanded cars that could cope with extreme dust, huge bumps and being parked all day outside a small, isolated pub with a corrugated metal roof. That's why Ford and Holden engineered their own cars for God's own country. Unfortunately, there aren't that many Australians in the world, and half of them are currently working in a pub in West London, which makes the business case for bespoke cars with tough chassis and big V8s hard to get past the accountants. So now Australia doesn't design or build its own stuff and has to make do with cars from Europe and America. This is not, in local terms, a 'no worries' situation. It's a 'some worries' situation. Although, in fairness, the average Australian probably has other things to worry about.

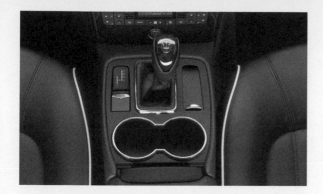

AUTOMATIC GEARBOX

A gearbox which changes gears automatically, hence the name.

AUTOTESTING

A motorsport in which a middle-aged man in an old Mini with the roof sawn off zooms about between cones in an old car park for no readily accountable reason. One of those motorsports it's probably more interesting to do than to watch. Although, come to think of it, that's pretty much all of them.

AXLE TRAMP

A loss of traction and juddering sensation caused by a poorly located axle on the driven wheels failing to contain vertical bounce from the tyres under high torque loads. Also the name of a German homeless man in Jeremy Clarkson's yet-to-be-written airport novel. Probably.

Look at me, I'm king of reversing a small, ruined car around some traffic cones.

B

BUGATTI

BMW

BOXER ENGINE

BRISTOL CARS

BRITISH LEYLAND

BARN FIND

BENTLEY

BEACH BUGGY

BEADED SEAT COVER

B IS FOR ...

BARN FIND

An old car discovered in a large farm building, often covered in straw and dust and hen poo. In bygone times this would have meant that the old car was ruined and must be thrown away, but that was before the classic car market went completely crackers. Now, 'barn find' still means a car that's covered in hay and fertiliser and owl pellets, but also one that for some inexplicable reason is worth about 15 per cent more than any otherwise identical old tat. Therefore, if you're trying to flog some rusty old Jag or Alfa that you just can't get shot of, why not sprinkle it with silage and goose mucus, add a few grand to the price and hey presto! It'll be sold in no time!

Look, I'm covered in sileage and pigeon dung. That'll be another £10,000, please!

BATTERY

Large block of lead and acid without which a car will not work. Unless you have left your car for a few days or weeks, in which case all the electricity will have somehow escaped and your car will not start, just when you really need it to. To sum up: car batteries are massive, heavy and they don't do their job at the most inconvenient of times, causing your entire car to stop working. Basically, they're the most under-achieving bit of car design by a mile. Sort it out.

BEACH BUGGY

Doorless and largely roofless fibreglass car based on a VW Beetle, originating in California in the 1960s and forever associated with a sort of care-free, wild-spirited grooviness that very rapidly evaporates if you have to drive one, say, hundreds of miles across Namibia. Or indeed anywhere in the British Isles when it isn't sunny.

BEADED SEAT COVER

Uncomfortable-looking wooden ball-based car accessory very popular with minicab drivers in the early 2000s (see also, pulling up outside your house and hooting the horn; not knowing exactly where things are; being apparently oblivious to the check engine light being on).

Look out! Sweaty indignation coming through!

BICYCLE — — — — — — ➤

Two-wheeled, pedal-powered device popular amongst children. Also a delightful way to saunter about the countryside, meander to a village post office or completely lose your temper because you are a self-righteous middle-aged man in unflattering lycra and a helmet that makes you look like a pompous mushroom and you are furious with a bloke in a van because how DARE he call you out for cutting across a pavement, jumping a red light and then riding the wrong way up a one-way street. I'VE GOT THIS ON CAMERA YOU KNOW.

BIG BLOCK

A large-capacity V8 engine made by General Motors in sizes up to 9.4 litres. The kind of thing that makes Richard Hammond smile in his sleep.

B IS FOR BENTLEY

British motoring-car company founded by W.O. Bentley in 1919. The company's early history is dominated by the 'Bentley Boys', a gaggle of plucky toffs who used to zoom about in their huge supercharged machines doing mannish things like winning Le Mans and racing trains. Also, one of their number was called Woolf Barnato, which might be the coolest name in history (even though his real first name was the slightly less impressive 'Joel' and his mother was called, we're not making this up, Fanny Bees). On the one hand, the Bentley Boys story is exciting and interesting. On the other hand, it's one of those bits of history that people just will not shut up about and which a car company can milk for the rest of time, as if a fat Swiss dentist with a brand new GTC drop top has any resemblance whatsoever to a man called Woolf risking his life in a race against a steam train. So modern Bentleys have literally nothing to do with Sir 'Tim' Birkin and all of that gentleman racer stuff, but they are very nice. In fact, the Continental GT is a rare example of a car that just keeps getting nicer and nicer over its life to such an extent that by 2050 scientists believe its niceness will only be visible to dogs.

B IS FOR BENZ, KARL

German engineer, generally credited as the inventor of the motor car in 1885. It was in 1888, however, that his invention really came into its own when his wife, Bertha, borrowed the first car without his permission and used it to complete an unprecedented 180-kilometre round trip. Karl was furious about this, not because she had taken the car but because when she returned it she had moved the seat and messed up his driving position. Probably.

B IS FOR BMW

Bayerische Motoren Werke (Bay area motorboat wipes). German car-making concern founded in 1916 as an aeroplane engine manufacturer for a thing we won't talk about. Moved into car making in 1928 with a licensed version of the Austin Seven, then went back into plane engines in 1939 for reasons we'll gloss over, then abruptly stopped all

that in 1945, again due to things we won't get bogged down in here. Now they make cars again. And what smashing cars they were and are. In particular, the M1, the M3 and the M5. Basically, if a BMW is named after a British motorway, it's probably good. Apart from the original M6. That was disappointing. Just like the motorway.

B IS FOR ...

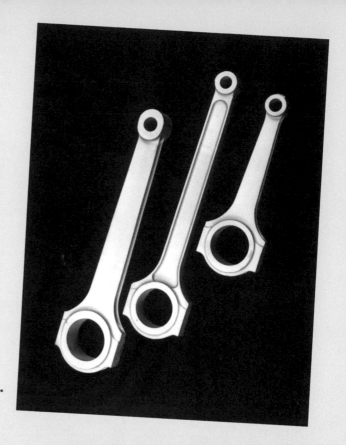

BIG END

The bottom part of the connecting rod between a piston and the crankshaft. Also a staple of unamusing 1970s comedies, e.g. 'Oh no Mr Longcock, me big end's gone and leaked all over Mrs Boobly's pussy!'

B ISTO ARS

Affectionate name for motoring car concern Bristol Cars, inspired by the constantly malfunctioning illuminated sign on their West London premises. It's a little-known fact that when B ISTO was owned by affable aviation tycoon Toby Silverton, the company would deliberately leave the sign with blown letters if it 'said something good'. Unfortunately, in 2017 the comically unreliable sign was removed and replaced with a boring logo, thereby signalling the end of civilisation as we know it.

BOOST GAUGE

A dashboard-mounted indicator of how much boost pressure a turbocharger is providing at any given moment. Of course, the times when a turbo is supplying any meaningful assistance are also the times when you probably should be concentrating on driving rather than staring at a small dial providing information of little to no overall importance. As such, the boost gauge exists at the very centre of a Venn diagram in which the two circles are marked 'coolness' and 'uselessness'.

BROWN SIGN

Poo-coloured road signs indicating the way to local tourist sites such as castles, museums and owlariums. An idea invented by the French in the 1970s and stolen by the British in the 1980s in order to make it easier to find the nearest owlodrome. Often left unusually vague (e.g. The Big Hole), perhaps in order to lure people towards the attraction in question and get them to pay an entry fee before realising how staggeringly dull it is (e.g. just a massive hole in the ground). On the other hand, it could be something really interesting like an owlitheatre. You just don't know.

BOXER ENGINE

A horizontally opposed engine in which pairs of pistons on opposing banks move in and out at the same time, which supposedly makes them look like a boxer's gloves if you really squint and don't know what a boxer's gloves look like. Not all 'flat' engines are boxers. The ones in Porsches and Subarus are. The one in the Ferrari BB 512 'Boxer', confusingly, wasn't because opposing pistons moved in opposite directions with each stroke, which made it, technically speaking, a 180-degree V12. TRUE FACT.

The very uninteresting bungalow

Museum of beige things

B IS FOR BRITISH LEYLAND

ARNOLD G WILSON WAKEFIELD LTD

British industrial conglomerate created in 1968 by the unholy merger of Leyland Motors and British Motor Holdings to form a foolish supergroup of car names that included Austin, Morris, Rover, Jaguar, MG and Triumph plus an unwieldy collection of other companies including Leyland (lorries and buses), SU (carburettors), Pressed Steel Fisher (stampings and bodyshells), Beans (foundries), Unipart (spares), Prestcold (fridges) and Nuffield Press (publishing).

Immediately began creaking under the weight of its own massiveness and insane duplication of effort which, along with an oil crisis and a turbulent time with industrial relations, caused the entire thing to collapse in 1975. Saved from death by nationalisation and soldiered on, cheapening Rover, neglecting Triumph, mishandling MG and general wiping snot on the family jewels of Britain's car industry until it was all ruined.

B IS FOR BUICK

The oldest-remaining American car maker, founded in 1899 with the sensational title of the Buick Auto-Vim and Power Company. Buick has an illustrious history, not only for being the founder of General Motors but also for being the first car company to fit an overhead valve engine, the first company to win at the famous Indianapolis Motor Speedway, the first company to fit turn signals and the first company to show off a concept car. Unfortunately, this once-innovative concern somehow lost its way during the 1970s and '80s and, despite occasional hotspots such as the scorchingly quick GNX coupe, largely became known as a provider of dull mush for old people. Would have almost certainly gone the way of Oldsmobile in one of GM's regular car company clearouts were it not for the sudden and surprise popularity of Buicks in China, where they are seen as attractive and interesting and not just a way for Aunt Luanne to drive to the early bird special at about 16 mph.

B IS FOR BUGATTI

French car company founded by an Italian in Germany. Early Bugattis were known for their exquisite design and fine performance but not for their brakes, which founder Ettore Bugatti considered to be of less importance. In essence, the history of Bugatti can be divided into three parts:

1909–1952 Made nice cars but then Ettore Bugatti died. Company went bust.

1987–1995 Made the cuckoo bonkers EB 110 but then there was a recession. Company went bust.

1998–present Came up with the Veyron and now the Chiron. Currently no plans to go bust.

C

CORVETTE

CONVERTIBLE

CHRYSLER

CAR BOOT SALE

CADILLAC

CAR DESIGNER

CAMPERVAN

CLOCK

CAR INSURANCE

C IS FOR CADILLAC

American luxury-car company, founded by Henry M. Leland in 1902 and named after – deep breath – Antoine Laumet de la Mothe, sieur de Cadillac, the French explorer who founded Detroit. Cadillac became known for innovations including the electric starter, the enclosed car body, the synchromesh gearbox, and indeed the layout of pedals and other controls we now take for granted. They also came up with the first mass-produced V8 engine and the first climate-control system. Cadillac became famous for lavish, exciting styling, culminating in the Eldorado of 1959, which had fins so big they could take your eye out. Unfortunately, after that Cadillac slowly slid into the GM mire of mediocrity, turning out massive but unremarkable saloons of the kind that keep rocking on their springs five minutes after you've parked them and are seemingly only bought by people with silver or bluc hair. Signs of a revival recently, helped by modern styling and the impressively sporty CTS-V. Despite their failings, 'Cadillac' still means 'really good' in the US. As in, 'Don't worry, Mrs Flappenberger, this is the Cadillac of walk-in baths.'

C IS FOR ...

CAR ADVERTS

Glossy print, video and online commercials which are heavily regulated to prevent depictions of speed and/or dangerous driving but are not regulated against irksome depictions of implausibly good-looking people putting surfboards, mountain bikes and frisbees into the boots of SUVs rather than, as would be more realistic, attempting to cram a screaming child into the back or get a load of sick out of the rear door bins.

CAR ALARM (GENERALLY)

Quiet sentinel protecting your car from getting stolen. Not that anybody would really pay much attention if it went off. Probably because everyone over a certain age was rendered immune to car alarms in the 1980s when they went off literally all the time, usually for reasons utterly unconnected to theft such as because it was breezy or raining or Tuesday.

CAR ALARM (IN TV SHOWS)

Television police detectives are just like us, except for one thing. When they lock their cars with the remote, their cars go 'bip'. Why does this happen? When you become a maverick cop who's going to crack this damn case if it's the last thing you do, does your chief constable insist that your car is fitted with a security system that goes 'bip'? And, if so, isn't this a bit dangerous? For example:
Criminal: Wait, are you the police?
Undercover cop: No, of course not, I'm on your side, Liam, you know I am.
Criminal: Then why did your car just go 'bip'?

CAMPERVAN

Small commercial vehicle converted into a mobile home by the fitment of a kitchen that's too small to be practical and a table that's too small to be useable which later turns into a bed that's too small to be comfortable. Or, for the purchase price, you could stay for several months in a hotel where there's lots of room to move and you don't have to poo in a bucket.

CAR

Honestly, if you need to know what this is you should start with another book, probably one from the children's section.

CAR BOOT SALE

Weird British tradition in which people take all the broken tat they don't want in their house anymore and attempt to flog it to strangers from the boot of their car. Rendered pointless by eBay, although going online doesn't allow you to spend a Sunday morning meandering around a local rugby club car park wondering how someone with a Nissan Almera has decided that this is the time to get rid of an onyx candle holder, a CD rack and a broken printer.

CAR DESIGNER

Highly trained aesthete much given to wearing black clothes and a deliberately weird watch and who is responsible for the overall appearance of a motor vehicle. Designing a car is an extremely complicated job which involves knowledge of aerodynamics, ergonomics and global regulations and then applying that knowledge to a shape, stance and set of details that will be attractive, saleable and easy to mass produce. It's a lot to take in, which is why some of them decide to forget about the 'attractive' bit, e.g. SsangYong.

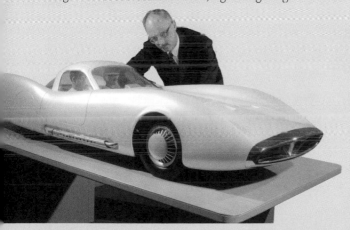

CAR INSURANCE

Legally mandated vehicle cover which appears to increase in cost every year even if you haven't used it because your car insurance provider presumably thinks that at some point in the previous 12 months you must have seen an accident or thought about an accident or... look, we know you can't be bothered to look around for anything cheaper so just give us an extra £114, okay?

CAR PORT

Seemingly pointless open-sided roof-on-legs very popular in 1970s Britain, as if cars were fine to get cold but must not get wet under any circumstances. Which, to be fair, in 1970s Britain was probably not far off the truth.

CAR SCATTER

Expression coined by James May to describe a situation familiar to anyone who owns two or more cars in which your vehicles all end up in different places, none of which are entirely convenient.

CARAVAN

Appalling fibreglass snail shell, the multiple purposes of which are to hold up other traffic while in motion, to snap off the back of your car when it falls over and to force you to defecate into a bucket like a medieval simpleton. Should have been rendered obsolete by civilised things like hotels but somehow soldier on thanks to weirdos who think a holiday should not be too far from home or indeed enjoyable.

CAR FARTS

Sturdy bum rattles released with regular abandon by anyone travelling alone in a motor vehicle and which will cause mild unease amongst someone who later has to get into the car as a passenger. Especially if the car is an Uber and they're paying for the pleasure of breathing in the ghost of last night's dinner.

C IS FOR CATERHAM

British sports car maker, founded after buying the design to the Lotus 7 in 1973 and then making it ever since. Originally from Caterham in Surrey, hence the name. Have spent 40-plus years fiddling about with the same basic design to great effect, such that nothing offers quite the same pure, intense, giddily windy driving experience. Tried something new with the Caterham 21 of 1994, which died after fewer than 50 had been made, and then again with the C120 project 2012, which died before it entered production. The moral of this story is, stick to what you know.

C IS FOR CITROËN

French car maker founded in 1919 by industrialist Andre Citroën or, as he is known in Britain, Andrew Lemon. As a car manufacturer, Citroën became famed for its innovative engineering, including adventures in the vanguard of front-wheel drive, monocoque body construction and aerodynamics, all of which reached a peak with the sensational DS of 1955 which also featured the hydropneumatic suspension that would become a Citroën signature for years. Unfortunately, while the company was very good at coming up with clever things like lights that swivelled with the steering, it wasn't quite so brilliant at making money and in 1976 the bankrupt business was sold to Peugeot, who then spent the following decades making their partner less and less expensively weird until they were largely selling bland hatchbacks under necessarily generous cashback deals. Still, all is not lost as the new generation of Citroëns is going back to ultra-soft (steel) suspension. And look, they have bumps on the side. So that's okay then.

C IS FOR ...

CARBON FIBRE

Strong, lightweight plastic reinforced with strands of carbon and used extensively in motorsport and high-end road cars. The use of fibres within a binding polymer results in a distinctive 'weave' effect in unpainted carbon fibre, the appearance of which can be copied and printed onto cheap plastic parts to make car interiors look totally disgusting.

CB RADIO

Short-wave radio frequencies open to all and very popular in the 1970s with lorry drivers and strange people in their back bedrooms who used it to recite numbers to each other for some reason. Fallen from popularity with the rise of the internet, mobile phones and dogging.

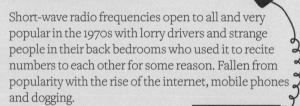

CHARACTER

Indefinable characteristic ascribed to some cars which makes them irrationally loveable despite, or perhaps because of, their flaws. Most often used in relation to Italian cars where it becomes a quicker way of saying, 'really good fun but one of the doors has fallen off'.

CHECK ENGINE LIGHT

Small orange catch-all which alerts the driver in a non-specific way that there is something amiss with their car and they should have it looked at immediately. Unless you are a British minicab driver, in which case the check engine light is something to be ignored for up to seven years. Not to be confused with the Czech engine light, which is only found in Skodas.

CHINESE CARS

Once derided for bring extremely ugly rip-offs of Western designs and called strange things like the Hung Sung 88X Mystery Vole, Chinese cars have suddenly become a lot more convincing in design and engineering and not being made of cardboard. Indeed, a Chinese company now owns Volvo, Lotus and the firm that makes British black cabs. In other words, laughing at the Chinese car industry is a bit like laughing at the crocodile that's coming to eat you.

CIGARETTE LIGHTER

What old people call a car's mobile phone and vape charger.

CLOCK

Car-mounted device for telling the time, most famously in Rolls-Royces where, according to *The Motor* magazine, at 60 mph it was the loudest thing you could hear. Sometime later it was observed that, rather than the motor car being very quiet, perhaps the clock was very loud.

COLLISION DAMAGE WAIVER

Optional car insurance policy available on hire cars which prompts the renter to have a moment of indecision while filling out the 412 forms required to rent a car. On the one hand, it's another expense and really, law of averages, what's the worst that could happen? On the other hand, without it you won't feel comfortable driving a hire car in the way it has to be driven, i.e. flat out everywhere and changing gear without using the clutch.

CONTINUOUSLY VARIABLE TRANSMISSION

Step-free automatic transmission which, on the plus side, allows the engine to hold at its most efficient speed whilst the transmission alters ratio to permit acceleration and, on the down side, makes the car go 'MUUUUUUUUUUUUUUUUUUUUOOOOOUUUUUUUU' in a way that's not very pleasant, especially when fitted to an old Volvo 340, which was also not very pleasant in general.

CONVERTIBLE

A car on which the roof can retract in some manner, thereby converting it from a warm, dry, comfortable means of transport into a draughty, dusty, surprisingly hot/cold (delete depending in climate) endurance test which only gets worse as speed rises and the entire experience becomes a tortuous maelstrom of wind and bees. Not popular in very sunny countries where everyone realises it's better to remain covered whenever possible. Hugely popular in Britain*, where it's often drizzling and where, on one of the four sunny days of the summer, many convertible owners can be seen driving around with the roof up as if they've forgotten that the retraction facility exists.

*Britain buys more convertibles than any other European country except Germany. TRUE FACT.

CORVETTE

'America's sports car', created by Chevrolet although these days it's one of those car names that they've tried to make a thing in its own right. Introduced in 1953, originally as a concept car and named after the smallest class of warship. TRUE FACT. Over the years the Corvette has been a vivid icon of Americana, from the '60s Stingrays of the early astronauts to the '70s C3 driven by packed-trousered porn star Dirk Diggler in *Boogie Nights*, all of which has allowed everyone to ignore the fact that many Corvettes weren't actually very good. But they are now, so that's okay.

CROSS SPOKES

Style of alloy wheel popular in the 1980s, notable for both looking excellent and being literally impossible to keep clean.

CRUISE CONTROL

Automatic speed holding system, the use of which can be extremely relaxing on long, straight open roads or, with appearance of other traffic, bends and/or a sudden realisation that this is an unfamiliar car and you can't remember where all the controls are, unbelievably scary.

C IS FOR CHEVROLET

American car maker, co-founded in 1911 by Swiss engineer Louis Chevrolet. Famed for its 'bowtie' logo which, legend has it, Monsieur Chevrolet copied from the pattern on some wallpaper. Aside from turning out thousands of boring, functional cars, the company has an illustrious history of doing interesting things that either work out well (e.g. the Corvette, the Camaro) or that are a total disaster (e.g. the Corvair, the Vega). In modern America Chevrolet is most famous for making adverts starring, as it keeps pointing out, 'real people' and which prove that, in fact, 'real people' are quite 'annoying'.

C IS FOR CHRYSLER

This year…you can easily afford a Chrysler!

merican car giant founded by Walter Chrysler in 1925 and now part of Fiat. Chrysler has achieved many things over the years and, though anyone who ever drove the 1995 Sebring Convertible might assume that includes 'making the worst car in the world', not all of its landmarks have been soggy rental car crap. They came up with the Airflow of 1934, which was one of the first American cars to use streamlining, they have a whole building named after them in New York even though they haven't lived there for years, and they used to make air raid sirens powered by their own Hemi engines. Imagine that, a V8-powered siren. It's enough to make you forgive them for the 1995 Sebring Convertible.

D

IS FOR ...

DAEWOO

DODGE

DAIMLER

DIESEL DACIA

DAIHATSU

DOUBLE DECKER BUS

DVLA

DRIVING INSTRUCTOR

D IS FOR DACIA

Romanian car company, named after the ancient region now occupied by its homeland. Founded in 1966 and generally busied itself making Renaults under licence until 1999 when it was bought by actual Renault who reheated a load of old Clio bits into the Dacia Logan and later, to the delight of James May, the Sandero and the Duster. Modern Dacias are a riot of clever cost-saving (all the door glass is the same across different models; the rubbing strips down the side are identical on both sides so they only need one mould to make; the left-side door mirror is the same as the right-side one, just flipped over; stop us if you're bored yet) and have a tough sort of feeling that comes from simple engineering and raised suspension to deal with the worst Romanian roads. However, there is also a simple rule that the more expensive the Dacia, the less appealing it somehow becomes. Also, despite what the ads say, you'll sound like a berk outside of Romania if you keep calling them 'Dat-cha', unless you're the kind of English person who insists the capital of France is 'Paree'.

D IS FOR DAEWOO

South Korean industrial conglomerate which clearly decided that car making sounded like a lark and, in 1982, bought the country's very first car company, Saenara Motors, which had been around since 1937, largely assembling other people's vehicles under licence. Decided to carry on in this vein by reheating various GM models until the mid-'90s when they came up with some equally underwhelming cars of their own, all of which were such a success that the entire Daewoo Group got into financial trouble and sold the car division to GM, which continued the great tradition of using the South Korean factories to make extremely dismal cars, sometimes badged as Chevrolets until they belatedly realised this was, like almost every part of this story, a terrible idea.

D IS FOR DAF

Dutch lorry maker that started making cars in 1958, using a strange system called Variomatic which basically employed rubber bands instead of a gearbox to effect some kind of continuously variable transmission. Bought by Volvo in the 1970s and slowly turned into their Netherlands branch office with responsibility for making their least-impressive cars.

AG·06·86

D IS FOR DAIHATSU

Japanese car maker, founded in 1951 and largely famous for its range of very small and slightly amusing cars, in particular the Charade GTti and the microscopically nutty Cuore Avanzato TR-XX R4. On reflection, it's entirely possible that some of the keys on Daihatsu's computer keyboard were sticking.

D IS FOR DAIMLER

1. German car company founded in 1890 and merged with the firm run by Mr Benz in 1926 to make cars badged, confusingly, Mercedes-Benz although the parent company was, and still is, called Daimler-Benz, except when it was Daimler-Chrysler. Is that clear? No, thought not.

2. The Daimler Motor Company of Coventry, not founded by Gottlieb Daimler although they did license his name, engines and cars from him. A Daimler found unfortunate fame when, in 1899, it became the first car to become involved in a fatal accident. Bought by Jaguar in 1960 and gradually reduced to fancier, ripplier-grilled versions of Jags and that large, stately limousine once much favoured by funeral directors and minor royals. Now officially 'resting'.

D IS FOR DATSUN

What Nissan used to be called before they changed the name to ... oh you know. Was dead, but has now come back for some reason, though only if you live in India, Indonesia, Nepal, Sri Lanka, Belarus, Kazakhstan, Russia, Lebanon or South Africa. Don't look at us, we don't make the rules.

D IS FOR DODGE

American car and truck company, founded in 1900 by the Dodge brothers as a supplier of parts to other car makers. Started building its own cars in 1914, as did 145 other companies that year, yet from that vast roster only Dodge survives to this day. Quite the CV in between then and now too, what with the Charger and the Challenger and the Viper and they're still doing mad things that light up their back tyres with the comically powerful Challenger Hellcat and, for people who think 707 horsepower isn't enough, the 840bhp Challenger Demon.

D IS FOR ...

DE DION-BOUTON

French company founded in 1883 by Jules-Albert de Dion, Georges Bouton and Charles Trépardoux and which was, briefly, the largest car maker in the world. Began as a manufacturer of steam cars and when they turned to internal combustion Monsieur Trépardoux, being a keen steam enthusiast, quit the company in a huff. It left the car business in 1932 and, after continuing to sell lorries, disappeared entirely in the early '50s. So, you know, maybe he was right.

DETAILING

What boring people on the internet say when they mean 'cleaning my car'.

DIESEL

Compression ignition internal combustion engine invented by German engineer Rudolph Diesel who disappeared in mysterious circumstances after boarding a ferry in Belgium bound for England. Given that his invention later became the noisy, soot-belching motor of choice for cheapskates and proved to be most useful for gassing monkeys, it's entirely possible he had great foresight and simply threw himself over the side.

DOOR POCKETS

A funny thing when you think about it. Door pockets. Door. Pockets. Doors don't have wallets and keys. The Lamborghini Countach has door pockets, even though its doors open upwards, causing everything in those pockets to fall out.

DOUBLE DECKER BUS

Bi-level public transport, the dual role of which is to a) provide greater passenger accommodation within a given vehicle footprint and b) get jammed under local bridges.

DOWNFORCE

Air resistance pressing on a car increasing vertical pressure on the tyres to the benefit of cornering grip. Explains the vast and complicated aero arrays on the front and back of racing cars and the intricate splitters and wings on hardcore road cars like the Porsche 911 GT2 RS and Jaguar XE SV Project 8. You know that spoiler on the back of your mate's BMW 320d? Yea, that's nothing to do with downforce, despite what he says.

DRIVING INSTRUCTOR

Unnervingly calm person sitting on the wrong side of a car being operated by someone who isn't qualified to do so.

DRAG COEFFICIENT

Measure of a car's resistance to air and therefore how slippery it is on the move, seized upon by marketing departments in the 1980s after Audi launched the 'aero' 100 in 1982 and put 'Cd 0.30' stickers in the side windows even though no one really knew what that meant.

DRIVING SHOES

Narrow-soled, flexible footwear designed to promote maximum grip and feel on car pedals. Ideal if you're a racing driver, not so necessary if you're just a tubby IT consultant doing a track day, not least because they make you walk like a delirious pigeon.

DUAL ZONE CLIMATE CONTROL

Double thermostat climate control system which permits the driver and passenger sides of a car interior to enjoy different temperatures and which, as a result, has saved more marriages than counselling and Ann Summers put together.

DRIVING GLOVES

Hand garments designed to give greater purchase on the wheel before the invention of a) power steering and b) social conventions which realised that wearing them makes you look like a bit of a wazzock.

DVLA

Driver and Vehicle Licensing Agency. British governmental body responsible for issuing driving licences and vehicle paperwork. Based in Swansea and, judging by their remarkably minimalist address, quite possibly the only business in town.

D IS FOR DELOREAN (CAR)

Stainless-steel-bodied, gullwing-doored coupe called the DMC-12, created by John DeLorean (see opposite) in an attempt to embarrass his former employers in Detroit with their yearly re-designs and inefficient, short-lived cars. DeLorean's 'ethical sportscar' wouldn't rust, wouldn't be facelifted every year and wouldn't run some huge, fuel-snuffling V8 engine. The prototype looked very nice, as it should since it was designed by Italian car-styling legend Giorgetto Giugiaro, but it didn't work so Lotus were hired to fix it, which they did by installing a modified Esprit chassis, hamstrung only by DeLorean's choice of engine, which was the unlovely Peugeot–Renault–Volvo V6, and by his insistence that it be slung out behind the back axle so there was enough room behind the front seats for golf clubs. DeLorean then managed to get a tidy government grant to build a brand new factory in Belfast, bang slap between Protestant and Catholic areas, and ignored protests from senior managers that it was hard to get work done when you looked out of the window and noticed your company car was on fire again. All told, the DMC-12 project started with noble aims and ended with its creator getting busted in an elaborate FBI drugs sting. Ironically, the car then became eternally famous as the star of the *Back to the Future* movies, a role it was chosen for partly because it looked cool but mostly because when the first film was made the DeLorean was an unloved comical underdog. And there's a reason for that: the sad truth is that the DMC-12 is and always has been a bit crap.

D IS FOR DELOREAN (MAN)

JOHN ZACHARY DELOREAN. American engineer
credited with inventing, amongst other things,
the first king of muscle cars, the Pontiac GTO. Was
also a lantern-jawed maverick who hung around with
Hollywood stars, decided to give Detroit a kick up the
bumhole with an incredible gullwinged sports car that
would expose their addiction to planned obsolescence,
rinsed the British government for a load of cash, got
into financial trouble with his factory in the turbulent
Belfast of the 1980s, attempted to save his skin by
brokering a multi-million-dollar cocaine deal, got
busted for it in a hotel near Los Angeles airport when
the whole thing turned out to be an FBI sting and got
off charges on an entrapment defence. But some other
parts of his life were probably quite interesting.

E

IS FOR ...

EXCELLENT

EDSEL

ECNALUBMA

ESPRIT

ELECTRIC
TAILGATE

EXHAUST

ELECTRIC CARS

E IS FOR ...

ECNALUBMA

What the front of an ambulance is called.

ELECTRIC TAILGATE

Powered hatchback operation, thereby motorising something that people seemed perfectly capable of doing by hand. Most systems sweep open with gusto, giving you an exciting opportunity to get smashed in the face by your own tailgate, and then close at the speed of an elderly slug, giving your dog ample opportunity to escape from the car and run into some fast-moving traffic. Yet strangely, not one car maker markets this infuriating and needless gadget as 'automatic jaw breaker' or 'fully powered dog murderer'.

ELECTRIC CARS

An increasingly popular type of vehicle in which the engine is replaced by an electric motor, the fuel tank is replaced by a battery and the liquid fuel in that tank is replaced by electricity, which is not a liquid and, according to James May, might not exist at all. Electric cars have actually been around since the 19th century but didn't gain favour because there was nowhere to plug them in. Now becoming more popular thanks to their winning combination of smoothness, quietness, torqueyness and not-giving-children-in-cities-asthmaness. On the downside, still quite hard to charge up because every time you turn up to a public charging station there's a bloody Mitsubishi Outlander PHEV plugged into it. You've got a petrol engine as well, you cheapskate lover of terrible cars, now PISS OFF.

EXCELLENT, THE

Unholy melding of a Land Rover Discovery chassis to a classic Mercedes SL bodyshell to create something that is a work of genius (according to Jeremy Clarkson) or sodding idiotic and pointless (according to Richard Hammond, James May and indeed everyone else in the world). After starring in the first series of *The Grand Tour*, The Excellent languished in a storage facility for a bit before Jeremy remembered that he'd invented it and brought it back to life. He now uses it to get around his Oxfordshire farm. Assuming it's not raining. And he doesn't really need to get to wherever he's going.

EXHAUST PIPES

Small tube permitting spent gases from an internal combustion engine to be funnelled away from the passenger compartment and safely allowed to disperse from the rear of the car. At least, that's what exhaust pipes used to be. Now they're like some deranged penis-waggling exercise in which the sportiness of a car is gauged by the quantity and diameter of pipework poking out of the back bumper. Some of them aren't even real because the actual exhaust points at the floor, which surely counts as the automotive version of shoving an old sock down your pants.

E IS FOR EDSEL

Attempt by Ford in America to create a slightly upmarket sub-company, famously such a total disaster that the entire endeavour only lasted from 1957 to 1960. The Edsel tanked because it was launched into an economic slump, because no one actually wanted a slightly upmarket Ford (and if they did they could have bought a Mercury) and, most importantly, because the front of an Edsel looked a bit like a lady's underparts. Ford rapidly became so desperate to flog these tragic-looking cars that they bought 1,000 ponies and dispatched them to Edsel dealers, the logic being that a pony on the premises would cause children to drag their weary parents into the showroom, whereupon they could be sold a car. Unfortunately, the average car dealer isn't well equipped to deal with live animals and the average buyer isn't keen to do a deal when all they can smell is poo. Also, although each dealer ran a competition to win the pony, many lucky winners went with the alternative prize, which was $200 in cash. In the end, hundreds of ponies were shipped back to Ford in Detroit, where they created both a mountain of horse crap and an accurate metaphor for the whole Edsel project.

E IS FOR ESPRIT, LOTUS

None-more-wedgy Norfolkian supercar, much favoured by Roger Moore-era James Bond for underwater and ski-resort shenanigans. Grew a turbo, got a V8, sprouted wings and things, had one major re-jig and stayed in production for a staggering 28 years. Yes, it was a bit homemade and slightly wonky around the edges, but to this day if you ask a child to draw a supercar, what they'll come up with, probably inadvertently, is an Esprit. This is what a supercar should look like.

PPW 306R

IS FOR ...

FIVE-SPEED GEARBOX

FLOOR IT

FUSE

FIAT

FAST LANE

FERRARI | **FSO**

FENDER | FORD

FUEL INJECTION

F IS FOR ...

FACELIFT

Word used by car nerds to describe the revised appearance of an existing model, usually released halfway through its lifetime. A slightly weird term to employ, since no car has ever had its metalwork pulled back and pinned behind its ears so that it looks fresher but also unable to look more than very slightly surprised. Still, it's not as odd as what BMW call a mid-term facelift. They call it a 'Life Cycle Impulse'. Perhaps that makes more sense in German.

FAP

Name given by Peugeot to the particulate filter fitted to their diesel engines. It stands for *filtre à particules*. Unfortunately, FAP also means something else. Something a bit more biological. Come on, don't pretend you don't know. That's right, it's also short for familial adenomatous polyposis, which is a rare genetic pre-disposition to developing cancer. Also, it means wanking.

FASCIA

Another word for dashboard, and one used exclusively by motoring journalists. No one in the real world actually says 'fascia', because you'd sound like a weirdo. 'Oh no, I've spilt some Fanta on my fascia'; 'I bought my Audi because it's got a really nice fascia'; 'Just pop this parking ticket on the fascia, would you?' See? You'd sound like a loony.

FAST LANE

What normal people call the outside lane of a motorway to the tutting disappointment of police officers, who insist it is known as 'Lane 3'.

FENDER

American word for the part of a car the British call a 'wing'. In the US they presumably rejected this word to avoid any confusion with the bits that stick out of the side of an aeroplane. Not that they were so picky with the whole gas/gas confusion.

FIBREGLASS

Fibre-reinforced plastic using strands of glass that can be quickly and relatively cheaply moulded into complicated shapes, thereby making it the ideal material from which to construct the bodyshell of a badly made British sports car.

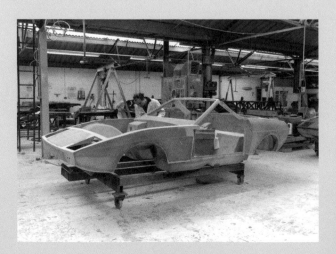

FIRE ENGINE

Large lorry carrying hoses, ladders, buckets, axes and people in burly clothes, primarily for the reaching and extinguishing of fires. It's very easy to regard the fire engine as too large, too unwieldly and too low of performance to do its job as well as it could, but try re-designing one and you'll soon realise that, unless you want your next fire to be dealt with by a bloke in a vest carrying two fire blankets and a bucket of water in a V8-powered Bedford Rascal, there's a good reason fire engines are the way they are.

FIRST TO SEE WILL BUY

Excessively cocky boast much seen in private ads for second-hand cars. Although, technically speaking, the first person to see the car after such a brag has been written is probably the person who wrote the ad and, as such, if the power of the car is as stated then they will have been forced to buy their own car. Either that or maybe, just maybe, they're talking bollocks.

FIVE-CYLINDER

Relatively unusual engine configuration, made most famous by the original Audi Quattro. Never a particularly popular format since it manages to combine the smoothness of a four-cylinder with the size of a straight six, which is exactly the wrong way round. However, five-cylinder engines are strangely delightful and give anyone describing them the perfect excuse to use the word 'warbling'.

FIVE-SPEED GEARBOX

Vehicle transmission containing a quintet of ratios, the very mention of which sounded almost insanely futuristic in about 1969, and now sounds almost comically lame and old-fashioned because most manual cars have six speeds and automatics have spiralled so madly out of control that some now have ten gears. If we plot this increase on a chart we can clearly see that by 2069 the average car will have a 24-speed gearbox, although James May has just pointed out that by 2069 most cars will be electric and the number of gears they will contain is one. So that's ironic, or something.

FLANGE

A flat surface used to attach or secure one item to another. For example, the protruding external body joins on an original Mini are flanges. Stop giggling.

FLAPPY-PADDLE GEARBOX

Casual way by which to refer to an automated manual or automatic gearbox operated by small flippers triggering electronic switches, usually mounted on the steering wheel. This expression was first coined by Richard Hammond on a television programme, the name of which escapes us for the moment.

FLATSHIFT

To change gear in a manual transmission car while keeping the accelerator firmly mashed into the carpet. A style of shifting used exclusively by racing drivers and people in hire cars.

F IS FOR FERRARI

Italian supercar maker and racing team founded in 1947 by former Alfa Romeo employee Enzo Ferrari, who only made road cars to fund his race team. And it's a good job he did, because without that reluctant money-raising gambit we wouldn't have the Daytona or the 512M or the 458 Speciale, and 1980s TV detective Magnum would have had to drive the Porsche 928 the producers had in mind in the first place (TRUE FACT!). These days road cars don't really pay for the racing, although the branded baseball caps, polo shirts and ironing-board covers probably do. Ferrari's Formula 1 team are famed for having periods of success and (much longer) eras of not doing as well or, more recently, times when they come perilously close to victory and then fall short and become extremely huffy about this, causing them to start stamping around as if everyone else is being unfair for not letting them win. And then they threaten to leave the sport but don't. It's basically an F1 team run by toddlers.

F IS FOR FIAT

Fabbrica Italiana Automobili Torino (Italian Automobile Factory Turin). Large car-making concern that was founded in 1899 and became Italy's largest car maker soon after and forever more. Notable achievements include mobilising Italy with the Topolino of 1936 and then keeping them rolling with the 500 of 1957, inventing the MPV with the 600 Multipla of 1956, providing affordable transport for the whole of rural Italy (and James May in Colombia) with the 1983 Panda 4x4, and building that famous factory with the test track on the roof. Over the years, Fiat became pre-eminent at making brilliant, modern small cars like the 127, Uno and Punto, which was fortunate as their big cars were generally hopeless, none more so than the Argenta saloon, which unfortunately went on sale in Britain just as the country went to war with Argentina and was notably unpopular as a result. Sadly, Fiat today seem obsessed with making all their cars look like the 500, apart from the 124 Spider, which looks like the old 124 Spider. Basically, they're a company selling many things that look like the stuff they used to make when they were more popular, like a middle-aged person tragically reverting to the hairstyle they had when they were young.

F IS FOR ...

FLATSPOT

A dead patch in the ongoing torque delivery of an internal-combustion engine, largely eradicated by modern engine-management computers, thereby robbing men of the opportunity to suck air in through their teeth while muttering, 'S'got a flatspot.'

FLEET MANAGER

Person working within a large business who carries with them the power to decide which employees are allowed which company cars. Basically, it's the power of life or death. Or, to be more exact, the power of Audi A4 S Line or Vauxhall Mokka 1.4 Active.

FLOODED

An engine that has been filled with an excessively rich fuel/air mixture and will not start as a result. The kind of thing your dad might have said many years ago. Doesn't happen now because of computer-controlled fuel injection, so if your dad wants to keep saying it he'll have to buy a classic car. They go wrong all the time. For some people that's part of the appeal. 'Some people' as in 'weirdos who like dismantling things rather than, for example, driving'.

FLOOR IT

Exhortation to depress the accelerator to its fullest extent. Often the preface to an accident.

FLYING CARS

Elusive combination of road driving and airborne ability that feels like it has been promised forever and yet which never arrives in any actual, practical form that you can buy. There is a reason for this and it's quite simple: a great many people are sodding terrible at driving a car when it's stuck to the earth, but at least with a land-bound car they can always slam on the brakes if things get a bit out of hand. Can you imagine if you put them in the sky? It wouldn't end well. Oh no, you say, you'd need a pilot's licence to operate a flying car and that's probably true, but then who's going to go to the bother, and without enough people how would you cover the massive development costs of making a truly commercial flying car for the mass market, eh? And where are you going to land the thing? On an airfield, that's where. And once you've done that, you might as well just park it and get into a proper car because flying cars are always a compromise and, since you don't want to muck about with the bit that keeps you suspended thousands of feet above the ground, the compromise always happens on the car part. Ergo, you end up with a crappy car that has wings strapped to the top. We already have cars and they're great. We already have a range of planes, and they're fantastic too. Plus, the big ones are already operated by highly qualified and competent people who do all the hard work for you. Come on, people! Stop dreaming that we can combine the two. It's like wishing for a pet that has the strength and great memory of an elephant combined with the beauty and easy containment of a tropical fish. Not going to happen. Unless we can develop reliable, affordable autonomy that can be applied to a workable car with flying capability. Then it might. So there.

FLYING LADY

The bonnet ornament on a Rolls-Royce motoring car, also known as 'The Spirit of Ecstasy'. The company says that the Flying Lady was based on an actress called Eleanor Thornton, but fails to mention that Eleanor wasn't her real name. Her real name was Nelly. Strangely, they don't seem keen to refer to their mascot as 'The Nelly', which is a shame.

FOG LIGHTS

Rarely necessary super-bright back lights, the deployment of which in mild drizzle makes it much easier to spot terrible drivers.

FORMER FAMOUS OWNER

Boast often seen in classic car adverts, presumably because it makes a car more interesting if it's been owned by a celebrity. But if you think about it, a rock star or actor is the last person you'd want to have owned a car before you. They've probably had lots of sex in it and then crashed it into a trout lake. And do you think Keith Richards or Oliver Reed was out there in all weathers scrupulously checking the tyre pressures and peering at the dipstick? Not likely. Really, these ads should crow on about cars that were once owned by a retired driving instructor you've never heard of and have in big letters at the top, 'Don't worry, NEVER owned by a famous person.'

FORMULA 1

Often referred to as 'the top level of motorsport' because no other car racing gives you quite as much technology, glamour and joyless processions around bleak new-build industrial estates while you slump on the sofa wondering if there's something more constructive you could be doing with your Sunday afternoons. For years F1 has wrestled internally over the conflict between its self-appointed role as a 'high-tech' sport and its obvious problem that if it leaves technology unchecked, someone will come up with a 3,000-horsepower, four-wheel-drive missile driven by a robot and there'll be even less chance of a fair fight on the track. Unfortunately, it's a wrestling match with itself that it still seems to lose. Basically, F1 has got two choices: either stop pretending to be sophisticated and go back to basics with simple, narrow, 1960s-style cars in which the driver does all the work. Or just admit that all bets are off, anything goes, and hope that at least one of the teams goes mad and turns up with something amusing like a rocket-powered monster truck. Until then, it'll be ugly cars driving round in circles and drivers talking in joyless monotones, as if being an international top-level racing driver is a right bloody chore.

FOUR-WHEEL DRIVE

System for sending engine output to all of a car's wheels (assuming it only has four of them) and which was once the preserve of off-road machines such as the Jeep and Land Rover before someone had the bright idea of applying it to sporty road cars. If you think that someone was Audi, lose ten points. If you think that someone was Jensen, it's a good effort but also lose ten points. The first four-wheel-drive 'sports' car was a Spyker, made back in 1903. It was also the first car with a straight-six engine, but let's not get bogged down in that now.

FOUR-WHEEL STEERING

System by which the rear as well as the front wheels turn to help a car change direction although by much less of an angle, unless you're one of those funny forklifts that can zoom about in tight spaces. Four-wheel steering was briefly exciting during the 1980s thanks to Honda and Mazda, then went away again only to quietly return on the Nissan GT-R and on various Porsches, Lamborghinis, Ferraris, BMWs and Renaults. Aids low-speed manoeuvring, boosts high-speed agility and, according to Jeremy Clarkson, makes passengers feel sick.

FRAMELESS DOOR

Feature of many convertible and coupe cars, and even some saloons and hatchbacks (looking at you, Subaru), in which the door glass stands alone and without a metal surround. On the one hand, expensive and complicated to engineer since it needs careful sealing to avoid excess wind noise and draughts. On the other hand, looks really, really cool.

FREEWAY

What American people call a motorway.

F IS FOR FORD

Officially the Ford Motor Company, which is, quaintly, how they still answer their phones. Founded in 1903 by Henry Ford, who had previously established what became Cadillac, and using money from the people who would later found Dodge. This just goes to show that Detroit in the early 20th century was quite an incestuous place. Ford's early years are most notable for the Model T of 1910, which made the car accessible to many more Americans thanks to its low price, which in turn was thanks to the efficiency of the newfangled moving assembly line (though Ford weren't the first car company to try this – Oldsmobile got there first). The massive success of the Model T and the philosophy behind it set Ford on a path they broadly follow to this day, which is to say providing affordable, practical transport for normal people,

although a modern Fiesta is significantly nicer to drive than a Model T on account of having all the controls in the right place and none of them outside the car. Because they haven't lost sight of their company-of-the-people roots, Ford have a great history of making blue-collar hero cars, from the Mustang to the Focus RS and even the Ford GT, which is like a Ferrari that listens to a lot of Springsteen. Unlike their big rival GM, Ford have largely resisted the temptation to keep bingeing on other car makers over the years, buying only Lincoln in 1922 and a chunk of Mazda in 1974, and inventing Mercury and the disastrous Edsel sub-divisions in-house. In the late 20th century they went a bit nuts and bought Jaguar, Aston Martin, Volvo and Land Rover, before burping them out whole, as if they were a big blue whale.

FIESTA ST
CONCEPT

F IS FOR ...

FRONT-WHEEL DRIVE

Vehicle design in which the front wheels are driven. On the plus side, allows for more efficient packaging because the engine can be mounted crossways and there's no need for a bulky hump in the floor for the propshaft, so more of a car's length can be used for people and luggage. On the downside, asking the front wheels to do the driving and the steering means they're a bit busy, and the car isn't as nice to drive as one with rear-wheel drive. That said, lots of front-wheel-drive cars have been fantastic. Sometimes referred to as 'wrong-wheel drive' by bores on the internet who don't know what they're talking about and have clearly never had a go in, say, a Renaultsport Clio 182.

FUEL CELL

Electrochemical unit that reacts hydrogen with oxygen in order to create electricity that can then be used to power a car. Believed by many to be the future of propulsion, and when that 'many' includes Toyota, Honda and Hyundai, you wouldn't bet against it. Let down at the moment by a lack of places to fill up with hydrogen. But then once upon a time you had to wait until the chemist's shop was open in order to buy petrol, and humankind sorted that one out in the end.

FSO

Fabryka Samochodów Osobowych (which, in English, basically just means 'car factory'). Polish concern originally set up to build other people's designs, initially from Russia and then from Fiat. Got carried away in the 1970s and came up with the Polonez, which was their own bodyshell on an old Fiat chassis and which over the years was used as an empty vessel for a strange palette of engines including the Ford Pinto, the Peugeot XU diesel and the Rover K-series. Hooked up with Daewoo in the 1990s and started making their cars until 2011, when FSO ceased to be a car maker. Which on the one hand was a shame, but on the other hand wasn't, because their cars were crap.

FUEL-FILLER FLAP

Hinged cover for your car's juice hole.

FUSE

The simple and easily fixable thing you hope is the root cause of the massive electrical problems in your car. See also Alfa Romeo; Character.

FUEL INJECTION

System for delivering fuel into an internal combustion engine. Once sounded exotic and sophisticated – e.g. Ford Capri 2.8 injection – and now sounds so completely normal that it's not even worth mentioning – e.g. Ford Focus 1.0 Ecoboost.

FUEL LIGHT

Dashboard tell-tale that illuminates when a car is running low on fuel, the reaction to which is either 'OH MY GOD, WE'RE GOING TO RUN OUT OF FUEL IN THE FOREST AND GET EATEN BY OWLS' or 'Ah ha, my old adversary. Well, let's see how far we can get with you smiling at me, eh?' Depends what kind of person you are.

TRUE FACT: Fuel fillers are on different sides on different cars, usually to be closest to the kerb on whichever side of the road they drive in the car maker's country of origin as a hangover from a time when petrol was dispensed at the road side by a cheery man in a brown coat. So British and Japanese cars, being from places that drive on the left, had their fuel filler on the left. German, French and Italian cars, made by people who drive on the right, also had their fuel fillers on the right. Obviously, we have petrol-station forecourts now, but the historical flap position still exists on a great many cars.

G

GORDON
BENNETT
GOLF
GULLWING
GENERAL MOTORS
GTI
GREEN CROSS CODE
GPS
GRUNT

G IS FOR ...

GOLF, VOLKSWAGEN

Medium-sized German car notable for being, in general, the answer to everything. Want a practical car? Buy a basic Golf. Want a very practical car? Buy a Golf estate. Want a sporty car? Buy a Golf GTI. Want a really sporty car? Buy a Golf R. Want an economical car, and also really hate monkeys and want to gas them? Buy a Golf diesel. Want to drive across a muddy field? All right, don't buy a Golf. You'll need a Suzuki Jimny or a Land Rover Discovery or something. But are you sure you need to drive across the field? Basically, if you're not especially interested in cars and you just want something that's useful and looks okay and won't get you sneered at in any social situation, the answer is probably a Golf and if the answer isn't a Golf then maybe you need to re-think your question.

GAS

What Americans call petrol. Not to be confused with actual gas, which they also call gas. Possibly didn't think this one through.

GORDON BENNETT

Old-fashioned British exclamation of surprise using the name of James Gordon Bennett Jr (1841–1918), American newspaper magnate, yacht racer and motorsport enthusiast famed for his lively and interesting life, most notably getting so drunk at a party in New York held by his (soon to be ex-) fiancée's parents that he urinated in a fireplace and was so embarrassed that he literally left the country and went to hide in Paris for several years. It was for one of his

GATSO

Popular brand of speed camera invented by Dutch rally driver Maurice Gatsonides, originally as a way of measuring his own speed when practising. 'I am often caught by my own speed cameras,' he once said. 'Even I can't escape my own invention because I love speeding.' He was basically the mouse that invented the mouse trap.

Gordon Bennett Cup races that the British team first painted their cars green rather than their traditional red, white and blue, and it's from this that we get British Racing Green. So thanks, Gordon. You were more than just a horrible faux pas in front of the (not quite) in-laws.

GPS

Global positioning system. Network of satellites that permits geolocation across the earth's surface and hence the operation of navigation devices via which blithering twerps can accidentally drive into a canal or across a live firing range and then blame it on their sat-nav. GPS is also what Americans call sat-nav. Tomato/tomato.

GREEN CROSS CODE

Basic road-crossing rules for children. First deployed in Britain in the 1970s using two promotional characters: a squirrel, because they're notoriously good at taking their time to cross roads, and a large man in a tabard, played by the actor David Prowse who later went on to be Darth Vader in *Star Wars*. But didn't do his voice. James Earl Jones took care of that afterwards. David Prowse was from Bristol, and presumably the *Star Wars* producers decided against having their baddie striding around going, 'Alroight, Luke, moi bouy. Oi'm yer farrrrther.'

GREEN-LANING

Pastime of 4x4 enthusiasts that involves driving slowly down unmade public byways in the countryside. The kind of thing Richard Hammond would enjoy and Jeremy Clarkson wouldn't. James May might enjoy it, if someone else was driving and there was a pub at the end.

GRUNT

Old-fashioned word for the outright 'go' of a car. Really it sort of means torque, but torque that's been splashed in some awful 1970s aftershave that smells of pine cones and extremely tight shorts.

GTI

Badge typically applied to the most sporting model in the range. Originally stood for Grand Touring Injection, although that meaning has sort of been lost now. Generally found on hot hatchbacks, although some time in the late 1980s that memo was put in the bin, hence the Peugeot 505 GTI and Montego GTI estate.

GULLWING

Roof-hinged doors pioneered by the Mercedes 300SL and so called because when they're both open and you look at the car from the front they make it resemble a seagull (if you really squint and don't know what a seagull looks like). Since that original Merc of the 1950s, gullwing doors have appeared on several other models, most famously the DeLorean DMC-12 and the back of the Tesla Model X. In engineering terms, gullwing doors have several plus points, such as permitting passenger access around a deep central tub, allowing more upright screen pillars for better visibility and requiring less space to open in tight spots. However, their main and only real advantage is that they just look really, really cool.

G IS FOR GENERAL MOTORS

American car-making megalith, founded in 1908 by the owner of Buick and destined to start sucking in other car companies from then on. Since their inception, GM's constituent parts have included Buick, Cadillac, Cartercar, Chevrolet, Elmore, Ewing, GMC, Holden, Hummer, Lotus, McLaughlin, Oakland, Oldsmobile, Opel, Pontiac, Rainier, Rapid, Reliance, SAAB, Saturn, Vauxhall, Welch and Yellow Cab. On the one hand, soaking up huge numbers of other firms over the years made GM mighty and globe-straddling, with a power and reach unobtainable to each of those firms on their own. On the other hand, you might notice that many of those companies don't exist anymore, while others have been sold to people who've taken better care of them. This seems to be the curse of GM, which has frequently blithered about like a kind of American Leyland, squashing independent thinking and blanding out what made each business unique, until too many of them became pointless and died. At their greatest, GM have come up with Corvettes and Camaros and Lotus Carltons and innovations like the EV1 electric coupe (TRUE FACT: the only car in the company's history to be badged as a GM rather than one of the marques within it). At their worst, they're a bloated purveyor of rental-car slush that made an absolute arse of SAAB and still doesn't seem to know what to do with Cadillac.

H

IS FOR ...

HYUNDAI

HIGHWAY CODE

HOLDEN

HONDA

HYBRID

HATCHBACK

HUMMER

HIGH STREET

HEADRESTS

H IS FOR ...

HANDBRAKE TURN

Rapid and dramatic rotation of an automobile within a small space using the parking brake to impede rotation of the rear wheels. The deployment of a handbrake turn is scientifically proven to increase the attractiveness of a young man to members of the opposite sex by 3,000 per cent, although this finding is sometimes disputed by certain groups, e.g. all women, and also your dad if he found out what you'd been doing in his car.

HARD SHOULDER

Broad strip down the side of major multi-lane roads, the main purpose of which is to provide people with somewhere to stop in the event of an emergency. In the olden days this was when the engine of your Humber or Wolseley malfunctioned. Nowadays cars are much more reliable, and as a consequence the hard shoulder is almost exclusively the preserve of people with punctured tyres or urinating children.

HARRIS MANN

Genial car designer responsible for the Austin Allegro who has spent the last 45 years quietly waiting for the tittering to die down before producing his original sketch of the Allegro design and patiently explaining that it should have looked much nicer, if only the general buffoonery of British Leyland hadn't got in the way. What's less reported about Mann is that he also did the Austin Princess, which still looks modern today, and the Triumph TR7, which has a big curving line up the side, just like dozens of car designs now, and that his post-Leyland career saw him doing consultancy work for big-hitters like BMW. He's basically a genius.

HAT

Head garment, the wearing of which while driving is a frequent indicator of incompetence of different varieties depending on the actual hat worn. For example:
Trilby – too slow and not paying attention
Baseball cap – too fast and not paying attention
Massive cat head – cannot see where going.

HATCHBACK

Practical but short-tailed car with a big door on the back through which you can put stuff. Like many things in motoring, the hatchback was invented and refined by the French, starting with Citroën and a version of the Traction Avant called the 11CV Commerciale (1938), before the practical baton was passed to Renault, who popularised it with the Renault 4 (1961) and 16 (1965). However, the French were not the first to fit a hatchback with fold-down rear seats. Weirdly, that was Aston Martin.

HEADRESTS

Padded and usually adjustable section at the top of a car seat, designed to minimise whiplash injuries. Standard on all modern cars, unless you are a character in a TV show, in which case your headrests are often missing and your car looks ridiculous as a result. (See also Central locking that goes pip when you press the fob button; Leaving your car with the window wide open; Not locking your car.)

HEAD-UP DISPLAY

System in which speed and other useful information such as sat-nav directions is projected either onto a piece of glass or directly onto the windscreen. Anyone buying a car with a head-up display typically uses it for about an hour after collecting the car while pretending that they are a fighter pilot and then turns it off because actually it's a bit annoying.

HAZARD LIGHTS

All-indicators-activation button, used primarily by idiots in BMW X6s as a sort of cloak of righteousness that somehow allows them to park on the double yellow lines by the cash machine, but also by normal people to say cheers to someone who's just let them out of a junction (see also Van driver's thank you).

HEARSE

Large, extended-wheelbase vehicle designed to carry coffins and travel at a speed that undertakers (the people who bury the dead, not drivers who nip round the inside of another car) consider not to be disrespectful (i.e. 4 mph). Hearses are typically based on large executive saloons and turned into a sort of mega-estate by an outside company. Would actually be very practical as a family car, but driving a second-hand hearse might seem a bit weird, unless your entire family are goths. Sometimes referred to as a 'hurst', but only by idiots.

H IS FOR HOLDEN

Australian saddle maker turned car-body builder turned part of General Motors in order to become a car company in their own right, albeit one that largely re-mixed other GM cars for Aussie consumption. Not that this was a simple task, as demonstrated many years ago when a batch of Opel prototypes arrived in Australia to be tested on local roads and, after a few days in the Outback, simply snapped in half. So Holdens were like other GM cars, but much, much tougher. And when they were allowed to do their own thing, as with the Monaro coupes and the Aussie-designed VE Commodore of 2006, they showed they could do honest, red-meaty, Chevy V8-powered fun like no one else. Unfortunately, there's only about 24 million people in Australia and half of them have gone travelling, so the economic case for bespoke Aussie cars is pretty poor and, despite valiant efforts to sell Commodores to the UK and US, in 2017 the last Aussie-built Holden was made and their factory has now been demolished. From now on, Holdens really are just lightly re-hashed GM cars imported from elsewhere. In local terms, a very un-ripper turn of events.

H IS FOR HYUNDAI

Vast South Korean industrial concern that founded a car division in the late 1960s making Fords under licence and then decided to have a crack at its own model, which it did by paying Giugiaro to style the body, buying Mitsubishi technology for the engines and poaching some people from British Leyland. The end result was the 1975 Hyundai Pony, which combined all the high performance of a small Mitsubishi with all the engineering pizzazz of a Morris Marina. That's right, it was crap. But Hyundai, being a vast South Korean industrial concern, were not mucking about and kept at this car-making thing until today, when we have the i30 N, which is brilliant, thereby proving the power of persistence even when all looks as hopeless as the 1994 Hyundai Accent.

H IS FOR ...

HEATED SEAT

Electrically powered heating pad in the base of a car seat that removes the chill on cold days and allows the seat's occupant to enjoy the faint sense of having wet themselves. Also ideal on hot summer days when you can play that game of trying to turn on the other person's heated seat without them noticing and then waiting about 90 seconds until they sigh and call you a rude word.

HEMI

Name used by various Chrysler car engines, originally derived from their dome-topped hemispherical combustion chambers. These days cylinder design has got a bit more sophisticated but the Hemi name continues to be used, basically to denote a sort of good ol'-fashioned, kick-bottom V8 of the kind that makes Richard Hammond go a bit glassy eyed.

HIGH STREET

Archaic: British term for the road where all the shops are.
Contemporary: British term for the road where all the coffee shops are.

HIGHWAY CODE

Small book of British road rules and regulations first published in 1931 and consistently one of the country's best-selling publications. All British people buy it, look at it before taking their driving test and then never read it again.

HIRE CAR

Someone else's vehicle, borrowed in exchange for money and three hours of your time as you fill in all the paperwork at the hire-car desk, even though you spent two hours at home filling in all your details online. All hire cars must, by law, display at least two of the following characteristics:
- Dents to three exterior panels (minimum)
- Interior that smells of synthesised apples
- Large quantity of sand underneath floor mats
- Funny sticky bit on back seat
- Mysterious noise that might be coming from left front wheel
- Completely empty glovebox (because you can't be trusted not to steal the owners' manual for a car you do not own).

HILL CLIMB

Form of motorsport in which cars take it in turns to drive as quickly as possible from the bottom of a hill to the top. Like many forms of motorsport, a lot more exciting to do than to watch. One of the most famous courses in Britain is Prescott Speed Hill Climb, which is owned by the Bugatti Owners' Club (although they let other people use it, otherwise it wouldn't be very busy).

H IS FOR HONDA

Japanese car maker (among other things) founded by Soichiro Honda, an interesting chap who started life mending bicycles, became a car mechanic, founded a company that made piston rings, and once became so despondent at his company's failings that he took a year off to make and then drink plum brandy. However, Mr Honda was not easily beaten. In the 1950s he decided that his motorcycles should win the Isle of Man TT race, even though they were massively outgunned on power and his team lacked experience. And they succeeded. In the 1970s he decided that he would win American customers round to his funny little cars with their fuel-efficient engines, even though he was up against the mighty power of Detroit's big three. And he did. Even by the mad standards of the kind of people who founded car companies in the 20th century, Mr Honda was a particular kind of ballsy. Little wonder his firm came up with some sensational cars, from the zingy little S600 and fabulous CRX to the original NSX and various Type Rs. Soichiro Honda died in 1991, which might explain why some recent Hondas have been a bit disappointing. Perhaps everyone in the company needs to take a year off to make and then drink plum brandy.

H IS FOR ...

HONEY BADGER, THE

Nickname of racing driver Daniel Ricciardo, presumably because he's tenacious and not because when threatened he routinely attacks people in the testicles.

HOOD

What Americans call a bonnet. The kind that stops a car engine getting all wet, not the sort you might wear on your head at Easter.

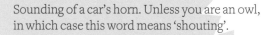

HOOTING

Sounding of a car's horn. Unless you are an owl, in which case this word means 'shouting'.

AMPLIFICATEURS DE SONS

HORSE BOX

Bespoke trailer created solely for the transportation of large equine animals, typically behind a Land Rover driven by a well-spoken person in a gilet who secretly thinks the contents of the box are more attractive and interesting than their spouse.

HOT HATCHBACK

Small-ish, versatile everyday car enlivened by extra horsepower, sportier suspension and a few tasty pieces of trim to create the perfect blend of practicality and driving amusement.

Many people think the first hot hatch was the Golf GTI. Actually it was the Simca 1100 Ti of 1974, which had more power, extra lights and alloy wheels, and went on sale over a year before the Golf.

HYBRID

Vehicle powertrain in which an internal combustion engine is combined with an electric motor for the purposes of greater economy or, in the case of the McLaren P1 or Porsche 918, brain-squelching acceleration. The modern mainstream hybrid was pioneered by Toyota, which is probably why it not only works properly but is able to stand the toughest mistreatment known to any car, which is to be a minicab.

HOVERCRAFT

All-terrain vehicle using powerful fans to create a 'cushion' of air between the underside and the ground, thereby allowing it to 'float' over rough ground and water, as long as the ground or the water isn't too rough and the driver isn't too fussy about being able to go exactly where they want to. The modern hovercraft was perfected by the British engineer Sir Christopher Cockerell and indeed the entire machine is somehow very British, being both ingenious and useful and yet silly and slightly ridiculous all at the same time.

HYDROPNEUMATIC

Suspension system invented by Citroën to cater for customers who wanted a car that had superior ride comfort and also started to resemble a shy turtle after it had been parked for a while. Ingenious for a number of reasons, one of them being an ability to drive even if you removed one of the wheels. This natural stability later saved the life of French president Charles de Gaulle after his hydropneumatically suspended Citroen DS was able to get away after being attacked by Algerian terrorists who shot out the tyres. Citroën has now ditched the hydropneumatic system because apparently it was too complicated and customers found it too weird, and presumably because they don't care if the president gets shot.

H IS FOR HUMMER

Defunct maker of off-road vehicles in a range of sizes from idiotically massive to idiotically sodding massive, all rooted in the decision to sell a civilian version of the US Army's Humvee military hack mainly because Arnold Schwarzenegger kept pestering them for one. GM bought Hummer in 1999, thinking this was a tremendous idea, then closed it down in 2010 because they were bankrupt and also because, as it turns out, it wasn't.

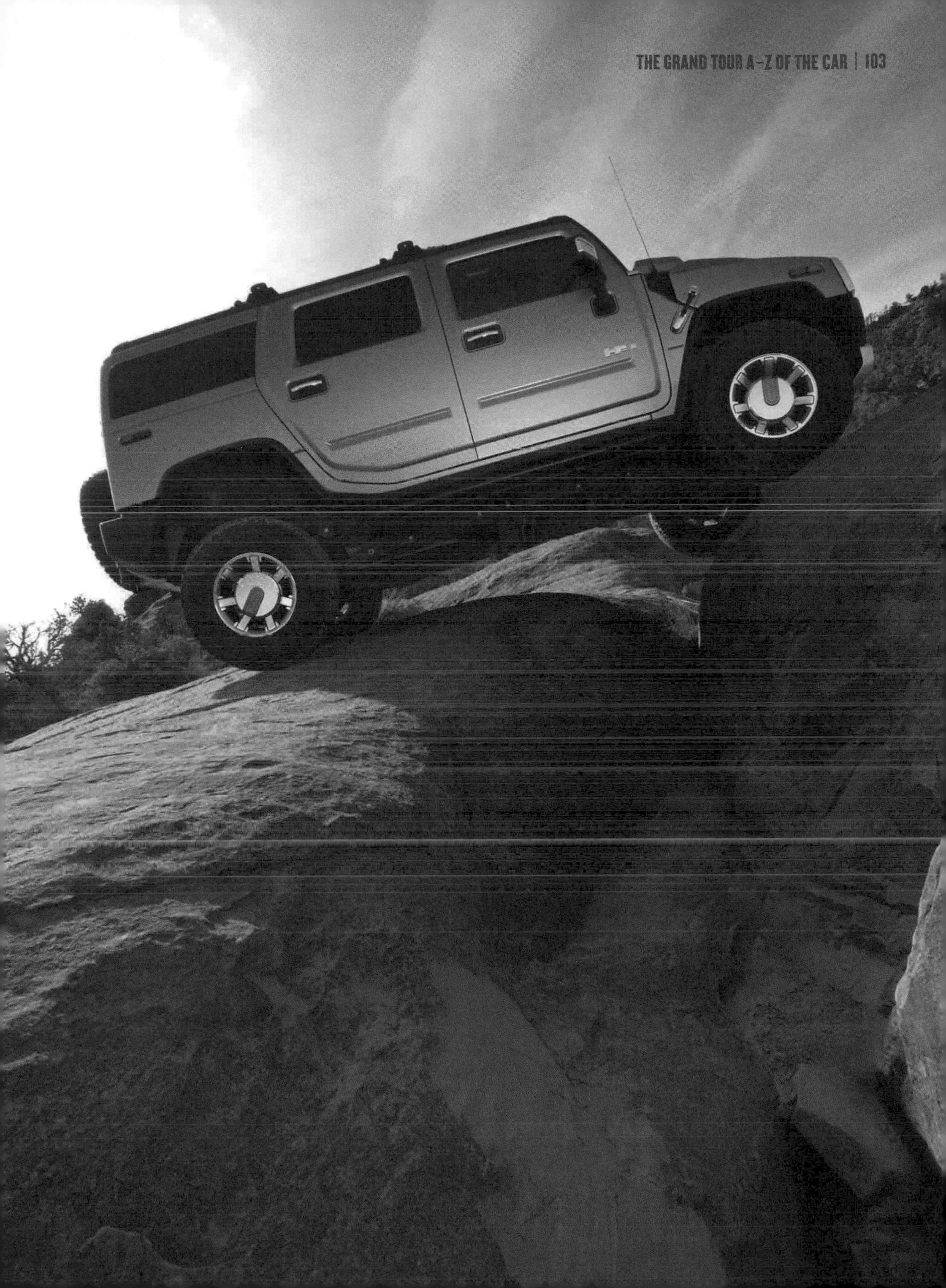

INDYCAR

ISUZU

ISLAND

INFINITI

ITALIAN

INDICATORS

I IS FOR ...

Regardez mon lampe d'indication!

INDICATORS

Flashing lights to indicate a driver's intention to turn, pull out or, if all four are deployed, briefly park on some double yellow lines so they can run to the cash machine. Introduced by Buick in 1939 as a replacement for previous indication systems, which were tiny trafficators that popped out of the sides of the car and/ or the driver's arm and were moderately useless on account of being tiny and/or only on one side of the car. Removed in the early 2000s by Audi on account of their customers having no call for them.

INDYCAR

American open-wheeled racing series, originally derived from Champ Car racing and with a name based on its most famous event, the Dycar 9000. Also the Indy 500, which is a 500-mile race around the Indianapolis Motor Speedway and which holds great lure for F1 drivers, although since tradition dictates that the winner must drink a bottle of milk, not the lactose-intolerant ones.

INFINITI

Relatively recent car company that is to Nissan as Lexus is to Toyota, i.e. slightly posher. For Americans, Infiniti arrived in 1989 with a promising car called the Q45, which looked a bit like what a Jag might be like if Jag had decided to be modern around then, which they hadn't. Didn't turn up in Europe until 2008, thereby refuting the expression 'Better late than never'. Continues to sell a range of forgettable things with unfathomable names and absolutely no selling points that would make you buy one over an Audi, Mercedes, BMW, Jaguar or Lexus because, as it turns out, very few people ever think that what they really want is a more expensive Nissan. On the plus side, Infiniti is a consistent contender for the title of the world's most pointless car company.

ISLAND

What people in the British West Midlands have decided to call a roundabout for no readily apparent reason.

ITALIAN

Latin-derived language spoken in Italy (and some parts of Switzerland and former Italian territories such as Ethiopia). Popularly regarded as a very passionate and musical tongue, and this is especially true when it comes to cars because many mundane car things sound better in Italian (as long as you don't actually speak Italian). A regular 1980s hot hatchback, for example. For Italians, it's just the Fiat Type Sixteenvalve. But for non-Italian speakers, it's the Fiat Tipo Sedicivalvole. *Bene!* Likewise, a 1980s supercar called, to Italians, the Ferrari Redhead. But for anyone not familiar with Italian, the Ferrari Testarossa. *Magnifico!* Everything car-related sounds better in Italian. *Olio* (oil). *Benzina* (petrol). *Incidente di parcheggio a bassa velocità* (low-speed parking accident). *Fantastico!*

I IS FOR ISUZU

Japanese car company, founded in 1916 in a joint venture between Tokyo Ishikawajima Ship-building and Engineering Co. Ltd, and Tokyo Gas and Electric Industrial Co., who decided to get together and make some Wolseleys. Had a glorious period when they built interesting things like the handsome Trooper 4x4 and the rakish Piazza coupe, then clearly decided this was all far too much and retreated to being a maker of commercial vehicles and diesel engines. Shame. The Piazza looked quite nice. The later ones had handling tuned by Lotus, you know.

J

JAGUAR

JEEP

JUMP J-TURN

JACKSON BROWNE

JENSEN

J IS FOR JAGUAR

British car maker, founded as the motorcycling-ist Swallow Sidecar Company in 1922, before becoming SS Cars and then, in 1945, after realising that wasn't a good name anymore, turning into Jaguar. Famed for fast, exciting, beautiful sports cars like the XK120, E-Type and F-Type, although it's in their saloons that the true heart of Jaguar is traditionally expressed, what with their powerful engines, their dynamic handling and their sense that the driver is probably going to steal all of your silver during the third week of his 'just one night' stay in your house, a move made necessary by what he terms 'a slight misunderstanding' with the mortgage company or his ex-wife.

J IS FOR JEEP

American 4x4 maker with their roots in the Second World War and the US Army's requirement for a small off-road machine to move things around and undertake light reconnaissance work. One of the companies contracted to make such a thing was Willys-Overland, who later trademarked the 'Jeep' name. They got bought by Kaiser, who got bought by American Motors, who got bought by Chrysler, who got bought by Fiat, and here we are today. Pending Fiat selling Jeep to Geely or something. It's never quite clear where the name 'Jeep' comes from, but the best guess is that it was coined by GIs and taken from Eugene the Jeep, a weird and unstoppable character in Popeye cartoons. One of those car names that's become a generic, like 'Transit' to mean 'van' or 'Prius' to mean 'minicab'.

J IS FOR ...

J-TURN

Dramatic reverse flick manoeuvre much enjoyed by no-nonsense heroes in movies, ex-SAS survival experts and normal people who want their car to start making a funny noise and then for one of the front wheels to fall off.

JACKSON BROWNE

To be extremely low on fuel. Inspired by the eponymous singer and his popular hit 'Running on Empty'. As in:
'Excuse me, mate, could you tell me where the nearest petrol station is? I'm Jackson Browne.'
'No, you're not. For one thing he's American.'
'Yea, cheers, you've been really helpful. Hey, wait. Aren't you James May off the TV?'

JUMP

To cause a car to briefly leave the ground, as seen a lot in rallying and 1980s American TV shows. Looks superb and, thanks to Hollywood, appears to be simple and perfectly normal, although in real life it's a tremendously good way to break all your suspension and/or make your car snap in half.

J IS FOR JENSEN

British car and lorry maker most famous for their Interceptor model of 1966, which was a high-water mark in the world of GT cars, not least for its simply enormous bonnet, its cavalier fuel consumption and its wonderful elegance, undermined only by the possibility that it might expire in a cloud of steam halfway between Geneva and Cap d'Antibes.

K

IS FOR ...

KITCHENETTE

KNOCK SENSOR

KEI CAR

KEYLESS

KIMI RAIKKONEN

KIT CAR

KIA

KOENIGSEGG

K IS FOR KEI CAR

Uniquely Japanese style of tiny car, built to comply with local regulations that date back to the late 1940s when war-ravaged Japan wanted to encourage small, economical cars to make best use of its limited resources. Over the years, kei car rules have been amended to make the cars bigger – no, really – but they're still constrained by the rules that say they can be no longer than 340cm, no wider than 148cm, and with an engine no bigger than 660cc making no more than 63 horsepower. For Japanese people, merely a practical fact of life if you live in a high-density city. For people outside of Japan, one of those things about the country that looks brilliant and weird and hilarious all at the same time, and makes you wish you lived there.

Actual kei cars include the Honda Life Dunk, the Suzuki Hustler, the Honda That's, the Mazda Scrum Wagon, the Honda N-Box Slash, the Daihatsu Naked, the Mazda Carol Me Lady, the Suzuki Every Joypop and the Mitsubishi Minica Lettuce.

K IS FOR KIA

South Korean car maker with their roots in a 1940s steel tubing company, and you know what they say about making steel tubing: just keep on going, via bicycles, until you've made a car. So that's what they did. Sort of. In fact, Kia Motors are South Korea's oldest car maker. And they carried on making cars until 1981, when the new dictator of South Korea told them to stop because he only wanted them to make small lorries. Seems a little petty, but there you go. Anyway, they bounced back and started turning out Mazda designs, then came up with their own stuff, which was a bit crap, and then they went bust, only to bounce back again as part of Hyundai. And from all this we've ended up with the Kias of today, like the Ceed and the Stinger, which are basically excellent. So that's nice, and inspiration to steel-tubing makers everywhere.

K IS FOR ...

KEYLESS

Car system that permits the locking/ unlocking of doors and starting of the engine without removing the key from your pocket or bag. Hang on, what's that you say, the WHAT from your pocket or bag? So it's not strictly keyless at all then, is it? See also Wireless internet. If there weren't any wires, the thing with all those flashing lights that lives on the side table wouldn't work now, would it? Discussion of all this and more can be found in the new book, *Sorry, Can I Just Stop You There...?* by James May.

KIMI RAIKKONEN

Linguistically minimalist Finnish racing driver famed for his excellent pace, especially when it comes to the alleged consumption of vodka.

KIT CAR

Homemade vehicle of a type very popular in the 1970s and '80s, when many people decided that the ideal form of transport was a wobbly fibreglass sports car with an engine from a Cortina and the axles of a Marina haplessly bolted together in the garage over the many months preceding the day their wives left them.

KITCHENETTE

Name given to the cooking area in a caravan or motorhome, as distinct from the seating and sleeping areas, which are many, many millimetres away, and the lavatory, which is a bucket outside.

KNOCK SENSOR

Engine sensor that detects the vibrations associated with unwanted additional combustion of fuel– air mixture after the initial spark and feeds this information to the engine-management computer so that it can address the issue by retarding the ignition timing. Not to be confused with a knock, knock sensor, which detects when your dad is about to tell a low-quality joke.

KNOWLEDGE, THE

Immensely challenging test all London black-cab
drivers must take and pass before being allowed to
ply their trade and which basically requires them
to learn all the roads in the biggest city in Western
Europe. An incredible skill to have, especially when
you consider that most cabbies have to leave some
room in their brains for their unsolicited views on
futboh, Mrs Fatcher and a bit about immigrayshun that
makes you think about just getting out and walking the
rest of the way.

KOENIGSEGG

Swedish supercar maker, founded by un-hirsute
genius and one-time frozen chicken magnate Christian
von Koenigsegg. Since their foundation in 1994, the
company have become noisily pre-eminent in the
fields of bonkers four-digit horsepower and doors
that open in a way no one else has thought of. And in
supercar land, both of those things take some doing.

IS FOR ...

LANCIA

LADA

LINCOLN

LOTUS

LAND ROVER

LEAF SPRING

LOCKING DIFFERENTIAL

LIMITED EDITION

LEXUS

LAMBORGHINI

L IS FOR LADA

Popular name for the products of AvtoVAZ, a Russian car maker founded in the mid-1960s by the Soviet government after securing a technology deal with Fiat following successful lobbying by the Italian Communist Party. Younger readers may not realise that communism was a big thing back then, and also that we were all scared of it. In 1970 Lada started making a version of the Fiat 124 saloon and in 1974 the car went on sale in the UK, at which point everyone wondered if they needed to be so scared of communism because this model was notable for being badly made and crap. Nonetheless, the Lada sold quite well in Britain, where it was much enjoyed in areas like the People's Republic of South Yorkshire, and in Russia, where there wasn't much choice. In fact, the original Lada is one of the best-selling cars in history, as well as the bedrock of punchlines for lazy 1980s comedians. AvtoVAZ went on to expand the Lada range with the Niva 4x4, which was weirdly appealing, and the Samara hatchback, which wasn't. When they failed to meet new emissions regulations, Ladas disappeared from view in Western Europe and the company eventually got bought by Renault–Nissan to give them massively increased penetration into the Russian market. Not very communistical at all.

HRH 697P

L IS FOR LEXUS

Posho part of Toyota, founded in 1989 with a name that is sometimes claimed to be derived from their original mission, 'Luxury Export to the United States', although Lexus themselves say the name is just a name and doesn't mean anything at all, so there. Famed for incredible build quality and usual Toyota levels of reliability, plus a lingering sense that most of the cars are technically brilliant but saddled with a Mercs-for-Freemasons image. The one exception to this is the LFA supercar, which is blessed with amazing design and God's own V10, and is one of the most incredible machines ever made. Ideal for anyone who appreciates brilliant engineering, or wants to get to the Lodge in a hurry.

L IS FOR LAMBORGHINI

Italian supercar maker, founded by tractor magnate Ferruccio Lamborghini, not, as is commonly thought, because he was snubbed by Enzo Ferrari and wanted revenge, but because he was a long-time Ferrari customer and knew that many of the parts in their cars were the same as in his tractors but with an attractively massive mark-up attached, and he wanted some of that profitability action. Obviously, this didn't entirely work out as Lamborghini the car maker teetered on the brink of bankruptcy several times over the years, leading to several changes of ownership before the company was finally bought by Volkswagen Group, who have at last steadied the ship by blending Italian madness and flamboyance with Germanic reliability and desire to make the doors fit properly. *Sehr bene!*

L IS FOR LANCIA

Italian car maker founded in 1906 and once a great innovator, being the first with a monocoque shell and the first with independent suspension, and in the vanguard of other significant developments such as the V6 engine and an integrated electrical system, including electric starter. Over the years, Lancia continued to live on or around the cutting edge, fitting its cars with front-wheel drive and five-speed gearboxes years before most car makers discovered such things, and their history is a litany of gorgeous machinery and motorsport success for which most companies would gladly sandpaper their own eyeballs to have in their back story. Unfortunately, technical innovation doesn't always go hand in glove with making money, and in 1969 the ailing Lancia were bought by Fiat, who gradually squashed the innovation and, although the glorious Stratos, 037, Thema 8.32 and Delta Integrale all appeared on their watch, by the 1990s Lancia were basically a branch office making re-bodied Fiat Tipos with Alcantara seats. Worse still, in recent times Lancia withered to such an extent that they became a company selling just one car (the Fiat 500-based Ypsilon) in just one country (Italy), and that model is coming to the end of its life, which means, although Fiat isn't making a song and dance about it, that Lancia are effectively a dead man walking. There are plenty of sad stories in the car world, but this might be one of the most tragic. Sniff.

L IS FOR LAND ROVER

British maker of off-roadish vehicles, generally claimed to have been founded in 1948, although that's when the original Land Rover itself was introduced and not when the company were founded, since at that point the Land Rover was just another product from the Rover company. Land Rover didn't become a separate company until 1978 and didn't split from their car-making sister company until 2000. Corporate semantics aside, Land Rover were originally known as a maker of affable farm equipment that looked best when it was dented and mossy, and have latterly become popular as a provider of school-run workhorses in various levels of bling, depending on your personal level of taste and/or matrimony to a Russian.

L IS FOR ...

LEAF SPRING

Car-suspension system using an arc of strong but flexible material, usually metal, singularly or in layers, to create a springing effect. On the plus side, simple, cheap and able to take a lot of weight. On the downside, rough, bouncy and prone to problems like axle tramp. Thanks to their medieval crudeness, leaf springs were phased out of European cars in the 1970s and American cars in the 2010s.

LIMITED EDITION

Vague catch-all for a variant of a car that is produced in finite quantity, although this is a broad church. McLaren made just five LM versions of its F1, and this might be considered a 'limited edition' as a result. Conversely, it's stretching the edges of the word 'limited' when you get to, say, a Vauxhall Corsa Breeze with standard tilt/slide sunroof and they've knocked out 10,000 of them.

LIMITED SLIP DIFF

A differential in which the difference in speed between the two output shafts is restricted, thereby permitting greater traction on split-friction surfaces and off the line in cars with a solid rear axle, and allowing easier performance of extremely massive powerslides for the amusement of television viewers if you are, for example, a 1980s television policeman or a 2010s *Grand Tour* presenter.

LOCKING DIFFERENTIAL

Function often seen in hardcore off-road 4x4s that allows the driver to stop the two outputs of a differential from turning at different speeds, thereby preventing a wheel or axle with low traction from spinning away all the engine's power. The kind of thing you'd shout if you wanted to get the immediate and undivided attention of Richard Hammond.

L IS FOR LINCOLN

American luxury-car company, named after President Abraham Lincoln and set up, confusingly, by the man who founded Cadillac as his sort of follow-up project once he'd sold his original car business to GM. Lincoln started in 1917 making aeroplane engines, then moved into cars and were bought by Ford, who wanted them as their upmarket division, something Lincoln could happily do with their fancy styling and V12 engines. In fact, Lincoln became a byword for American luxury, culminating in sensational, elegant machinery like the Continentals of the 1960s. In fact, John F. Kennedy himself was riding in a Lincoln when he was shot and killed, something that might have been bad for business, though not as bad for business as Ford's cack-fisted management of the company into the 21st century, when they began flogging milky re-hashes of ordinary Fords, slathered with extra chrome and rumpled leather, and an unconvincing saloon built from the same box of parts as the unloved Jaguar S-Type. More recently hit the comeback trail by getting Matthew McConaughey to mumble things in their TV ads. Oh, and by coming up with some new cars. Although most Americans just remember the Matthew McConaughey bit.

L IS FOR LOTUS

Sports-car company and racing team from Norfolk, although actually established in London – TRUE FACT! – by company founder Colin Chapman. Known as an innovator on the race track, with monocoque chassis, wings, ground effects and active suspension, as well as being the first team to take commercial sponsorship; also a trailblazer in road cars, with a lightweight philosophy, fibreglass monocoques, turbocharging and experiments in how much people would pay for a plastic car with visibly wonky panel gaps. In recent times, the company's motorsport efforts became trapped in a confusing situation in which two different F1 teams were claiming to be Lotus, even though, strictly speaking, neither of them were, while their road cars have largely been endless rehashes of the Elise chassis that first came out in 1996. In modern times, as always, if you want the most wonderful steering and the best handling in the world, get a Lotus. If you want a study in solid build quality, you're probably more of a Porsche person.

GDD 85L

AU09 AKN

M

IS FOR ...

MINI

MORRIS

MOTORWAY

MERCEDES-BENZ

MASERATI

MERCURY

MG

MAYBACH

MINICAB

McLAREN

M IS FOR MASERATI

Italian sporty car maker founded by Alfieri Maserati and his three brothers, all called Maserati. Their first names were different, obviously. Otherwise that would have been really confusing at home. Achieved great success in racing, became the only Italian team ever to win the Indy 500 (which they did twice), moved into road cars, gave up racing, gave us wonderful grand tourers like the Mistral, Ghibli and Bora while getting bounced around between various owners including, weirdly, Citroën (1968–75), who used a Maserati engine in their SM coupe of 1970 because clearly their own engines weren't powerful or unreliable enough. Weathered the 1980s making the boxy and rather unsexy Biturbo, and then got bought by Fiat, which is what happens to all Italian car companies (except Lamborghini). Twenty-first-century Maserati specialise in dated GTs and almost good-looking saloons, nothing is quite as good as it should be, especially the inevitable diesels and SUVs, and the whole company feels like no one really knows what to do with it.

M IS FOR MAYBACH

German car company that briefly made luxury cars between 1921 and 1940 before concentrating their attentions on tank engines, which Germany needed a lot of around this time for some reason. Bought in the 1960s by Mercedes, who kept ignoring the to-do list with 'Re-launch Maybach' on it until 2002, when they started making a massive and slightly vulgar limo based on the old 1990s S-Class and which seemed to exist only to give rap and R&B artists something to name-check in songs that was easier to make scan than Rolls-Royce. Since 'rappers and a few Russians' aren't really enough of a demographic to sustain a car company, Maybach disappeared in 2013, only to rise again as the trim level on a stretched, super-lux version of the Mercedes S-Class. P Diddy must have been delighted.

M IS FOR MODEL T, FORD

American automobile made from 1908 until 1927. Henry Ford is often misquoted as boasting in relation to the Model T that 'You can have any colour, as long as it's black,' which isn't quite what happened. For the first few years of its life the car was available in several colours, none of which were black, until, in the quest for greater efficiency, Ford decided to slash the number of permutations his assembly line turned out and reportedly said to his sales people, 'Any customer can have a car painted any colour that he wants so long as it is black.'

From this point in 1914 onwards, Model Ts only came in black, reportedly chosen because it was the paint that dried the fastest, which didn't matter because, as contemporary footage shows us, the whole world was black and white anyway, and also all cars were capable of cornering at 200 mph with 19 policemen hanging off the side. In 1926, towards the end of the Model T's life, they re-introduced some other colour options anyway.

M IS FOR MAZDA

Japanese car company founded in 1920 as Toyo Cork Kogyo Co. Ltd, which made cork until 1931, when, in a move that would get you thrown out of Dragons' Den because your business model was too diffuse, they came up with a three-wheeled motorcycle truck that they called the Mazda-go, named after Ahura Mazda, the god of harmony, intelligence and wisdom. Moved into making cars in 1960 with the R360 micro car, and have been at

it ever since. Mazda have a long history of doing unusual things, like successfully bringing back the small, light, two-seater roadster in the late '80s, when no one else could be bothered, and slightly nutty things, like persisting with Wankel rotary engines when everyone else had decided they were a bad idea. They are, on the quiet, the coolest Japanese car maker, especially now Honda have lost their way a bit.

If you've ever wondered why the Mazda logo is all lower case except for the D, it's because they wanted all the letters to line up perfectly to show the precision of their engineering and a lower case D would have spoilt that by poking up above the other letters. Not sure why they didn't just make everything upper case. Maybe they thought it would be too shouty.

M IS FOR McLAREN

British racing team and road-car maker founded in 1963 by Bruce McLaren, who was actually New Zealish. Enjoyed huge success in Formula 1, winning eight constructors' championships and 12 drivers' titles with distinctly unshabby talent such as Ayrton Senna, Niki Lauda, James Hunt, Alain Prost and Lewis Hamilton, and had an impressive record in Can-Am and at the Indy 500 and Le Mans. Enjoyed a couple of faltering starts at being a road-car maker, first with the M6GT of 1970, which never made it into production, and again with the F1 of 1992, which was wound up after just 64 road cars with no immediate successor. Fortunately, they started having a proper crack in 2011, starting with the MP4-12C and on to the 650S, the 675LT, the P1 and so on. Turns out they're really good at it, which is fortunate because at the moment their racing team is crap.

M IS FOR MERCEDES-BENZ

197 MBC

German car, van, bus and lorry company co-founded by Karl Benz and therefore able to lay claim to be the inventors of the car, which probably doesn't do any harm, marketing-wise. The other founder of the company was, of course, Gottlieb Daimler, and to this day the overall parent company of M-B are called Daimler-Benz, but Austrian car salesman and keen Daimler fan Emil Jellinek famously suggested their cars would sell better if named after his daughter, Mercedes, and the rest is history. So it's a good job she wasn't called Bertha, although funnily enough Mr Benz's wife was. It's from Daimler that the company got their famous three-pointed-star logo, the triple tips supposedly representing land, sea and air, and therefore the company's interests, which also extended to making boats and engines suitable for aeroplanes. Daimler and Benz merged in 1926, and went on to create a whole slew of legendary cars including the 300SL Gullwing, the 300 SEL 6.9, the G-Wagen, the 600 'Grosser' (strictly, the 'Großer') and of course the Unimog. Everyone loves a Unimog. Used to specialise in a limited range of very understated saloons and sports cars, but nowadays have a range so massive they're in danger of running out of letters with which to name them all.

Mercedes racing cars became known as the Silver Arrows after they scraped all their normal white paint off to save weight, leaving a bare aluminium shell.

M IS FOR MG

Formerly British car company founded by a Morris employee called Cecil Kimber in 1923 for the purposes of making two-seater sports cars. Became famous for such things through various spindly wheeled, vintage-looking machines like the M-type, TC, TD and the original TF, although they also made plenty of MG-badged saloons, and took part in all manner of motorsport, including the Le Mans 24-hour race and, bizarrely, NASCAR. However, as always with British sports-car makers, MG's history is quite a saga and one that, inevitably, involves getting sucked into the British Leyland monolith, becoming starved of investment and then being allowed to die before unexpectedly springing back to life. Twice.

In MG's case, they became part of BMC in 1952 and part of British Leyland when BMC were merged with Leyland Motors in 1968. They then ceased to exist in 1980 when their two remaining models, the MGB and Midget, were killed off and the Abingdon factory closed, only to rise from the dead as the badge used on sporty Austins like the Metro and Maestro, then to fade away when these were discontinued, only to rise from the dead again with a bizarrely re-heated MGB running a Rover V8 called the MG RV8. It was a warm-up act for the proper return of an MG sports car with the MGF of 1995, which became a whole range of saloons and hatchbacks after BMW sold MG and Rover to some Brummie businessmen in 2000.

The company teetered on the brink of death in 2005 when the Brummie business project went south, only to rise from the dead AGAIN when the MG name was bought by a Chinese company called Nanjing. This was in turn absorbed into SAIC, who now make MG-badged hatchbacks and SUVs in China, and who make a half-hearted effort to pretend they're British because the company maintain a small engineering centre on the outskirts of Birmingham (and a really massive one in Shanghai). Cecil Kimber would probably be horrified, if he hadn't died in a train crash in 1945.

M IS FOR ...

MACPHERSON STRUT

Car suspension design based around wishbone at the bottom and a vertical strut acting on a coil spring and damper unit at the top. Invented by General Motors engineer, Earle S. MacPherson. Also a name for the way car engineers dance.

MEMORY SEATS

Useful facility that allows cars with electrically adjusted seats to return to the place where you left them after someone else has been driving, thereby avoiding that annoying sense when someone has used your car that you'll never quite get the seat back to the way that you like it, much the way your teeth feel like they don't quite fit right after an intense trip to the dentist.

MERCURY

American car maker invented out of thin air by Ford in 1938 to sell very, very slightly more upmarket companions to the products of the mothership. Managed to do this for decades, largely selling re-hashed Fords with more chrome and pleather, until they were killed off in 2010, to the dismay of almost no one.

METALLIC PAINT

Car paint containing tiny metal particles to give the surface a more pronounced shine. A useful way for car companies to get another £450 out of you.

MIDDLE LANE

The central lane of a three-lane road, where people who aren't very good at driving live. Permanently.

MILK FLOAT

Small, electrically powered lorry for the dispensation of dairy products and, if 1970s British films are to be believed, to permit the milkman to silently prowl the streets seeking sexual intercourse with other men's wives.

MINICAB

Private-hire taxi, traditionally a medium-sized saloon or hatchback with one scuffed panel, mismatched wheel trims, an unsettling sticky patch on the back seat, an overpowering smell of synthetic fruit almost but not quite masking an undercurrent of sweat, farts and some kind of meat, and driven by a man who doesn't entirely know where he's going but does know how to signal his arrival at your house, which is to stop outside someone else's house and then sound the horn. In many towns and cities this quaintly English tradition has now been replaced by the app-based stylings of a well-known American taxi company, and the driver doesn't need to not have a clue where he is going as the GPS system on his phone will now accurately guide him to slightly the wrong place.

MIRA

A top secret car-testing facility at which the most sensitive of highly classified prototypes can be unleashed in the maximum-security environment just off the A5 near Nuneaton between Higham on the Hill and Fenny Drayton at Watling Street, Nuneaton, CV10 0TU.

MINI (COMPANY)

A classic case of a car model that became a whole make (see also Land Rover). In the case of the Mini, even the Mini wasn't called the Mini at first. It was the Austin Seven or the Morris Mini Minor for the first 10 years of its life until, in 1969, Austin and Morris badges were removed and 'Mini' became a thing in its own right, although they were still sold through Austin, Morris and, later, Rover garages. Mini only became properly stand-alone in 2001, when BMW, having kept the name and the British-engineered new Mini that was in the works when they sold Rover, set up dedicated Mini showrooms and a dedicated Mini range of cars that are very amusing but aren't especially mini.

M IS FOR MINI (CAR)

ARCHAIC Microscopic city car launched in 1959 and made until 2000 in a variety of styles and specifications, becoming the very epitome of 'Swinging London' 1960s cool as well as, in Cooper trim, the cheeky, zippy template for all sporty small cars ever since. Surprisingly valuable now, even though most classic Mini owners seem to be infected with a kind of disease that makes them want to plaster their entire car with stick-on chrome tat, badly made wooden dashboards and a whole bucket of fake '60s crap, despite their car being a 1987 Mini City automatic formerly owned by a retired headmistress.

MODERN Range of German cars styled to look like a cartoon version of an original Mini that's developed elephantiasis and marketed with heavy emphasis on the sort of Austin Powers interpretation of the '60s, which people in other countries sometimes believe is an accurate depiction of Britain to this day. Latest affectation, installed on the smallest hatchback model, is rear lights that look like the Union Flag of the United Kingdom – a tremendous idea from Mini's design studio in Munich and looks great coming out of their factories, one of which is in the Netherlands. Pip pip! Cup of tea bulldog bowler-hat scurvy crumpets, old boy!

M IS FOR MITSUBISHI

Japanese car maker founded in 1917 when the Mitsubishi Shipbuilding Company decided to have a crack at being land-based and came up with something called the Model A, which would become – TRUE FACT! – Japan's first mass-produced car. From this strong start, Mitsubishi went on to some great things, like the Starion, the Galant VR4 and various flavours of Evo. They also innovated, being the first company to fit active suspension in production, and they were big fans of the balancer shaft, which is an ingenious way to make an engine smoother. Unfortunately they also came up with the awful 3000GT and invented a dull mush of a car that they called the Carisma, thereby redefining all known measurements of irony. Basically, Mitsubishi were once brilliant and now they've gone a bit rubbish. They can't even be bothered to replace their famous Shogun model, which is now about 100 years old, and after a scandal involving overstated fuel-economy claims they slid so far down the dumper they had to be bought up by Nissan, who will hopefully show them some photos of the excellent Galants of the '80s and '90s or an old Evo VII and remind them to stop being cack.

M IS FOR MORGAN

None-more-British motoring-car manufacturer founded in 1910 to make funny little wood-framed cars with Edwardian suspension and the permanent requirement to wear some kind of tweed hat. Since those early beginnings, Morgan have come a long way and now make funny little wood-framed cars with Edwardian suspension and the permanent requirement to wear some kind of tweed hat. Since they've managed this for over 100 years, they must be doing something right. Indeed, not long before his death in 1991, Soichiro Honda himself predicted that in future the world would contain just six car manufacturers. Having made this grand and accurate prophesy, he paused before adding, 'Oh, and Morgan.'

M IS FOR ...

MISLEADING BADGES

A bootlid applique from a car manufacturer with an alpha-numeric model-naming system that gives deliberately false information about the size of the engine. The rot started with the BMW 525e of the 1980s, which, contrary to BMW's badging policy of the time, actually had a 2.7-litre engine. This opened the floodgates of fibbing, so that today the BMW 340i has a 3-litre engine and the 330i is only a 2-litre, while Mercedes merrily badges a 2-litre car as the C300 and sells an E63 with a 4-litre V8. It's dishonest, and, when James May comes to power, it will become illegal.

MODEL CODE

Internal reference system used at all car companies to identify models and variants and eagerly seized upon by those of a nerdish persuasion to both show they know what they're talking about and to throw a geeky cloak of exclusion around their favourite subject. Hence BMW fans harping on about 'E30s' and 'F10s', or Merc fans waxing lyrical about 'W124s' and 'W140s'.

MODEL YEAR

Confusing label applied to cars, which doesn't tally whatsoever with actual years since model years seem to start in about August and sometimes earlier. Hence, a car launched in May 2019 could be described as a 2020 model year even though it's strange and nonsensical. Model years aren't much of a thing in most of the world but are a huge deal in the United States, presumably as a result of car makers' historic love of yearly updates as part of a sneaky programme of planned obsolescence, which is why the first question an American will ask you about a car will be along the lines of 'Is it an '18?', and woe betide you if you don't actually know because you're British, you don't really care that much and you just wanted to rent a Mustang while you were on holiday.

Ford used to do the best model codes. The mk3 Escort was codenamed 'Erika', replacing the mk2 which was 'Brenda'.

MONKEY GASSER

A Volkswagen diesel.

MOTORCYCLE

Two-wheeled vehicle that, according to Richard Hammond or James May, is available in a vast range of sizes and styles, contains fascinating precision engineering, and is symbolic of true freedom and liberation and the feeling of the wind in your face and, according to Jeremy Clarkson, has lots of different types, all with lots of Zs and Xs in their names, and all look the same and are stupid.

MOTORWAY

What British people call a freeway.

MOTORWAY SERVICE AREA

Dedicated facilities beside major arterial roads in which one may use the lavatory, buy some food, get stuck behind a slow-moving woman using a single crutch, observe a large shirtless man having a shave in the lavatories and buy a large model of a duck made out of scrap-metal pieces for no readily apparent reason. British motorway service areas are in many ways more lavish and sophisticated than those in other countries, yet they also manage to be seven times more depressing and inexplicably capable of making everyone in them look extremely whey-faced and unattractive. Our advice: if you just need a quick wee and a coffee, go straight to the petrol station bit rather than the main services. The bogs are often in a terrible state but it's much quicker.

MOTORWAYS, POLICE CARS ON

The world's most accurate benchmark for 68 mph.

MULTIPLEX

Wiring system in modern cars in which each electrical device within the bodyshell contains a tiny 10-screen cinema.

M IS FOR MORRIS

Defunct British car maker founded in 1912 as WRM Motors and based in the Cowley area of Oxford. Thanks to popular cars like the 'Bullnose' Morris, the company became very successful and by the mid-1920s had overtaken Ford to become Britain's largest car maker. Morris lost that title but continued to thrive on a reputation for providing quality cars at reasonable prices and in 1948 came up with the Morris Minor, which was designed by Alec Issigonis (later to get much more praise for the original Mini), and which later became Britain's leading source of things that go 'wu-oomph'. It was also hugely successful, becoming the first British car to sell over one million examples. In 1952 Morris merged with their great rivals, Austin, in an attempt to present a unified front against foreign rivals, but this only led to many Morrises becoming badge-engineered Austins and to the creation of British Leyland, neither of which were good things. Inevitably, Morris went into decline and the name finally died in 1985. As a final indignity, the very last Morris-badged machine was actually an Austin Metro van built not in Morris's traditional home of Cowley (where Minis are now made) but in the once-hated Austin factory at Longbridge. Worse still, in 1984 the last actual car to wear a Morris badge also came out of Longbridge and, even worse than that, it was a Morris Ital. Which is like realising that the one thing people remember about you before you left the party is that you vomited on the host.

Nº **2**
ASSY LINE

N

IS FOR ...

NISSAN

NIGHT PANEL

NOBLE

NACA DUCT

NIPPLE

NOTCHBACK

NASCAR

NEGATIVE CAMBER

N IS FOR ...

NACA DUCT

Distinctively shaped air intake so-named because it was first developed by the National Advisory Committee for Aeronautics (which later became NASA) and which aims to allow air in without disturbing the flow across the surface it's set into. Not brilliant if you need loads of air to enter your intake, but on the plus side they do look very cool and are found on many cars including the Lamborghini Countach, the Ferrari F40, the Dodge Viper, the Alfa Romeo Montreal, the Ruf CTR Yellowbird and the Nissan GT-R.

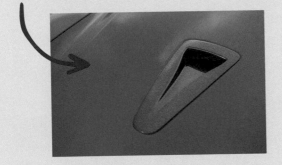

NEGATIVE CAMBER

Suspension set-up where the bottom of the car's wheel sits further out than the top, giving the wheel a slight 'lean' when static. Often seen on the rear wheels of BMW M cars and, weirdly, the old Vauxhall Insignia. The positives of negative camber are that in hard cornering it compensates for the forces on the wheel to maintain a larger contact patch with the road, and also that it looks cool.

NASCAR

National Association for Stock Car Auto Racing. Catch-all term for long-running stock-car racing series in the US. For many Americans, NASCAR is practically a religion, with a fascinating heritage harking back to the highly tuned cars bootleggers would use to outrun the authorities during Prohibition, an honest simplicity to the mechanical make-up of the cars, a roster of relatable heroes for drivers, a promise of close racing, serious speed and dramatic accidents, and an ability to draw six-figure crowds looking for a good time. For many people in other parts of the world, it's just inbreds driving in circles.

NOTCHBACK

Strange and archaic word for a car with a flat bit of bootlid at the back, but not a very long one. So a 1983 Jaguar XJ6 is emphatically not a notchback but a Mk III Ford Escort is. In case you were wondering.

NIGHT PANEL

Feature introduced by SAAB that allowed the driver to turn off all the instrument lighting except the speedometer. It was supposed to be less distracting at night and, like most things SAAB did, was claimed to be 'aircraft-inspired'. Also like most things SAAB did, it was copied by no one.

NIPPLE

A small metal fixture that allows the application of lubricant to a bearing using a special grease gun. Why are you giggling?

NOBLE

Small British sports-car company in the great tradition of small British sports-car companies, that is to say selling plastic cars using other people's engines and dashboard switches, loosely nailed together in a light industrial unit just off the M1. However, Noble's cars have always been more than that, chiefly thanks to electrifying performance from heavily tweaked Ford and Volvo engines and sensational handling that could take the fight to Porsche and Lotus. Sadly, the man responsible for much of this, founder Lee Noble, left the firm in 2008, and it's now run by a chap called Peter Boutwood, although he's wisely decided not to change the company name since Boutwood M600 sounds like a type of cricket bat.

N IS FOR NISSAN

Japan's first car company, founded in 1911 as Kaishinsha Motor Car Works and going through various name changes before ending up as Datsun and then, in the 1930s, Nissan (although in most countries the cars were still badged Datsun up until 1983). Nissan found their feet as a car maker building a facsimile of the Austin Seven and went on to create some terrific stuff, including the 240Z, the Sunny GTI-R, various Skyline GT-Rs, leading up to the incredible GT-R of today, the original Primera (which was so good – TRUE FACT! – it was Ford's benchmark when they were developing the first Mondeo) and, of course, the Serena diesel, which was for many years the slowest car on sale in Britain, with a 0–60 time of 26.5 seconds. Nissan were also the first Japanese car company to set up a factory in the UK, opening their Sunderland plant in 1986 to build the Bluebird, which wasn't an exciting car but turned out to be just as unswervingly reliable when made by Mackems as it was when it came from Japan. Of course, nobody's perfect, and Nissan has also come up with some total dross over the years, not least the Juke, which looks like a talking tumour and is guaranteed to send Richard Hammond into a rage at the mere sight of its stupid headlights.

Q

OVERSTEER

OWNERS' MANUAL

OIL

OPEL

OVERTAKING

OCTANE

OPEN GATE

OLDSMOBILE

OPEN-FACE HELMET

O IS FOR ...

OCTANE

Chemical component of petrol used as a rating based on knock resistance, the number of which is employed as a marketing tool to suggest the 'quality' or 'powerfulness' of the fuel. Also, a glossy magazine popular among people who go to the Goodwood Revival in a very strident pair of trousers.

OIL

Natural or synthesised engine lubricant and substance that you don't want to see on the ground under your car.

The perfect lover

Castrol GTX

The oil with protective instincts.

Castrol GTX. The Engine Protector.

OPEN GATE

Thing that really annoys farmers. Also the name for a style of gearshift often seen in Ferraris (and Lamborghinis, and indeed the original Audi R8) in which there is no leather gaiter at the bottom of the lever and a metal 'gate' indicates the different gear positions. Famed for making a nice clack noise as you change gear. And also for being a bit slower than a normal gear selector. Largely forgotten, now that most supercars have paddle shifts. Oh, well.

OPEN-FACE HELMET

Head-protection device lacking the horizontal bar that covers the mouth and chin. On the plus side, allows the rider of a motorcycle to feel more connected to the world around them and permits television presenters in safety-critical situations to talk to the camera without looking like their voices have been dubbed on afterwards. On the downside, makes it much easier to swallow a bee. Also, you look a bit of a chump.

The first Ferrari to have the open-gate gear shift was the famous 250 GTO of 1962. Oh, yes.

OVERSTEER

Motor-vehicle cornering characteristic in which the front wheels follow a tighter arc than expected, with the consequence of making the rear wheels swing wide. Terrifying if you're not ready for it and have no idea what to do with it. Absolutely hilarious if you are and you do.

OVERTAKING

The act of passing another vehicle. On the single-lane roads of the UK overtaking used to be commonplace and a good way of making progress, unimpeded by horse boxes and blithering slowcoaches. Today, to overtake a slow-moving car, even on a clear, well-sighted A road in fine weather with no oncoming traffic for seven miles, is tantamount to suggesting to the overtaken person that you will come round to their dormer bungalow and wipe your bottom on their commemorative Brexit tea towels. As a consequence they will become extremely angry, and as you go past they will quite probably do something significant like flash their lights at you.

OWNERS' MANUAL

Absolutely enormous publication contained within a plastic or fake leather wallet and which contains information about every single detail of a car in such excruciating detail that not one person on earth has ever read such a thing cover to cover nor indeed ever looked at it once.

O IS FOR OLDSMOBILE

Defunct American car company founded by the superbly named Ransom E. Olds in 1897 and bought by General Motors in 1908. Oldsmobile can claim to have pioneered the first mass-produced car, beating Ford to the idea of a production line, and were in the vanguard of early V8 development, coming up in 1966 with the world's most powerful front-wheel-drive car, the sensational Toronado with its 7-litre V8 engine, later upgraded to a staggering 7.5 litres. Oldsmobile also gave their cars brilliant names that sounded like fighter jets, including the Super 88, F-85, Starfire, Jetstar and Cutlass. Even their in-house V8 was called the Rocket. Unfortunately, in some kind of strange, self-fulfilling prophecy, Oldsmobile increasingly became seen as a retired person's car company, and this problem got so bad that in 1988 they tried to tackle it by mentioning it in their ads, which only seemed to make things worse. Oldsmobile died in 2004 as did, by the law of averages, several of their customers.

O IS FOR OPEL

German car company founded by Adam Opel in 1862, but to make sewing machines and bicycles because the car hadn't been invented yet and clearly Mr Opel couldn't be bothered to come up with that one. Got into the car business in 1899 and became known for affordable, dependable cars, including something called the Doktorwagen, a reliable, reasonably priced car aimed specifically at physicians making house calls. In the 1920s Opel experimented with rocket power and got bought by General Motors (although these two things are almost certainly unrelated), and in the 1930s became the first German company to sell a car with a monocoque body, built in a factory that could claim to be the most productive in Europe. By the 1970s and '80s Opel were very successful at making good workaday cars that were often a bit nicer than equivalent Fords, plus some delightful velour-lined smokers like the Monza and Senator. Then in the '90s Ford decided to stop phoning in lowest common denominator dross just as Opel got a bit slack, culminating in the miserable Vectra of 1995, and although their recent stuff has been better it hasn't been very profitable. In fact, GM Europe haven't made a profit once in the 21st century, which might be why in 2017 General Motors head office decided to flog the whole lot to Peugeot.

P

PARK & RIDE

PETROL STATION

PAGANI

PS

PCP

PETUNIA

PROTON

PROP SHAFT

PAVEMENT

P IS FOR PAGANI

Italian supercar maker founded by ex-Lamborghini engineer Horacio Pagani who started his own business in 1988 after his former employer didn't share his enthusiasm for new-fangled carbon composites. Pagani showed off its first car, the Zonda, in 1999. It was going to be called the Fangio in honour of Mr Pagani's friend, racing driver Juan Manuel Fangio, but he died in 1995 so they changed the name as a mark of respect. Since it's been around for just two decades, Pagani is something of a Johnny-come-lately in the high-end supercar world, but it just

goes to show that you can still turn up years after Ferrari and absolutely belt it out of the park with a V12 monster that looks and drives like nothing else. The Zonda was notorious because it was very good, because it came with a pair of driving shoes made by the same cobbler the Pope uses, and because every time they announced the absolute final edition they seemed to pause and then announce yet another one, making it the car they literally couldn't stop building. Who knows, perhaps they're still making it today.

You know who owns a Zonda? Lewis Hamilton. Is that interesting? I'm not sure.

EK-650VV

P IS FOR PROTON

Terrible car company founded by the Malaysian state in 1983 that launched their first car – a badly re-heated Mitsubishi – in 1985 to rapturous acclaim in the home market, largely because the government had specifically banned the media from saying anything negative about it. Continued on largely dismal form thereafter, although in 1996 they did buy Lotus, who did their best to make the cars feel less dreary and actually succeeded with the slightly amusing Satria GTi. Proton soldier on to this day, although they've been bought by Chinese giant Geely and were pulled out of Europe years ago to a huge wave of complete disinterest.

P IS FOR ...

PARK & RIDE

Urban-management system in which a motorist is encouraged not to drive their car to the place they wish to go, park, get out and go into the shop, restaurant or cinema they wish to visit, but is instead urged to leave their car several miles from their final destination and use an intermittent bus service to reach a place quite near the shop, restaurant or cinema they wish to visit. Not exactly a tempting option, all things considered.

PARKING SENSORS

Vehicle-manoeuvring assistant using ultrasonic sensors to detect and indicate the proximity of other objects. A nice idea, but in many cars the parking sensors are needlessly panicky and start beeping like a deranged robot when you could fit an entire fridge in the gap between your back bumper and the car behind. Ideally, such systems should have a range of settings that dictate when they start the really urgent beeping, ranging from neurotic (six inches) to London (one inch) to Paris (actually touching).

PAVEMENT

What Americans call the surface of a road and what a British person calls the slightly elevated pedestrian area at the side of a road. Ergo, if you're British be careful not to tell an American to 'walk on the pavement', or they will get run over.

PCP

Personal contract purchase. Finance arrangement usually based on an initial deposit followed by monthly payments over a set period, after which you can pay a lump sum to keep the car or simply hand it back. The way a great many people get cars these days, which is why car adverts no longer breathlessly tell you about performance or sex appeal and mostly just fill you in on the PCP arrangement, so that the average car ad on the radio basically sounds like a string of advanced calculus.

PETROL STATION

Thing that used to stand on the outskirts of your town, where those new flats are now.

PETUNIA

Codename of the 2004 Ford GT, chosen because it sounded delicate and gentle and would therefore throw anyone who heard it off the scent and prevent them from guessing that Ford were about to build a 550-horsepower, V8-powered supercar.

PICK-UP TRUCKS

Functional open-bed vehicles seen as essential workhorses in the developing world, lifestyle statements in the United States, and a bit moronic in Europe.

POLICE INTERCEPTOR

Slogan written on British police cars, presumably as a result of the following conversation:
'Hey, Sarge, can we have guns like American police do?'
'Certainly not.'
'Awww, but we want to look cool.'
'It's out of the question.'
'Fine. Can we at least write INTERCEPTOR on our cars?'
'Yeah, whatever, go nuts.'

PRINCESS ANNE

See Reliant Scimitar.

PROCON-TEN

Short for 'programmed contraction and tension', Procon-Ten was a car safety system announced by Audi in 1986 that used a thick steel cable wrapped around the back of the engine and gearbox and then connected via pulleys to the steering column and seatbelts so that in the event of a frontal accident the engine would be shunted backwards on a pre-set path, pulling on the cables, which would then snatch the steering wheel out of the driver's face and yank all the belts tight. In many ways it was ingenious. In other ways it was also heavy, complicated and expensive, and it smashed your steering wheel through your dashboard. Little wonder all other car makers decided airbags were a better idea and, by the mid-'90s, Audi agreed.

PROP SHAFT

Mechanical connection for transmitting torque, typically from the end of the gearbox to the differential of a rear-wheel-drive car, or from the engine to the rear-mounted gearbox in cars with a transaxle design, using a flexible connection such as a universal joint to account for the relative movements of the items at either end. Generally simple and reliable unless you are Jeremy Clarkson and you are driving an Alfa GTV in Scotland, in which case your prop shaft will be up to all manner of shenanigans.

PS

Pferdestärke (which is German for horsepower) and also known as metric horsepower, defined as the ability to raise 75 kg by one metre in one second. Used extensively by car makers in their badges (e.g. Ford Mondeo ST220, Jaguar XE 300 Sport), PS has fallen out of favour with boffins who talk in kilowatts, and is not to be confused with bhp (brake horsepower), which is very, very slightly different. Here's a tip: to convert PS to bhp simply multiply by 0.987. [That's quite enough now thank you, May – JC]

P IS FOR PERODUA

Perusahaan Otomobil Kedua Sendirian Berhad (which translates as Second Automobile Company). Malaysian company which, as their full name suggests, was set up to provide the country with another car maker after Proton. Well, if your only car company was Proton you'd probably try

again too. Perodua generally make lightly re-hashed Daihatsus and sometimes Toyotas, and are actually now bigger than Proton, which is good because Perodua's cars are quite amusing in an old-fashioned sort of way, whereas Proton's are cack.

P IS FOR PEUGEOT

French company founded in 1810 as a maker of bicycles, one of many pies Peugeot has had their digits in, some of which remain fingered to this day. Among its business interests, Peugeot has at various points made watch mechanisms, sewing machines, munitions, corsets, irons, tools, washing machines, motorcycles, radios, food processors and salt and pepper grinders. In 1889 they also decided to have a crack at cars, coming up with a short-lived machine powered by steam before switching to petrol power,

paving the way for the Peugeot we know today. In recent times Peugeots were rugged and upright in the 1960s and '70s, then became sporty and stylish in the 1980s and '90s (apart from the 309, which was meant to be a Talbot, perhaps explaining why it looked so gawky). Unfortunately, somewhere around the turn of the century, Peugeots lost their way, becoming mediocre to drive and often unpleasant to look at. They seem to be pulling it back a bit now – and don't worry, their pepper grinders have always been excellent.

P IS FOR PININFARINA

Italian design house, formerly written as Pinin Farina. Founded in Turin in 1930 by Battista Farina and part-named after his family nickname 'Pinin', which means 'littlest brother' in the local dialect. Pininfarina were originally a coachbuilder for companies such as Alfa Romeo, Lancia and even Rolls-Royce, and did work for Cadillac and Nash in the US. In 1951, however, they struck a deal with Ferrari and became best known as the designers of all the great Ferrari road cars that followed (except the 308 GT4 of the '70s, which Bertone styled). Pininfarina have also been responsible for some of the best-looking cars from other manufacturers, including the Alfa Spider, the Fiat 130 Coupe, the Peugeot 405, the Hyundai Matrix... actually, forget that last one. Anyway, they've styled a lot of nice-looking cars over the years. And the Hyundai Matrix. All has not been well at the company recently, however, as Ferrari have decided to design their own cars (with frankly mixed results), and Pininfarina have experienced financial troubles for some time until, in 2015, the whole company was bought out by the Mahindra Group.

It wasn't all Ferraris, you know. Pininfarina were the people who turned the MGB roadster into the GT coupe and the Jaguar XJ6 Series 2 into the Series 3.

P IS FOR PONTIAC

Deceased American car maker, founded in 1926 and named after the area of Detroit where Pontiac were based. For people outside the US, Pontiac are most famous for making the Firebird and in particular the uprated Trans Am version, which is best known, depending on how old you are, as 'that car from *Smokey and the Bandit*', 'that car from *Knight Rider*', or 'that car driven by that weird guy from the other side of town who wears vests and a cowboy hat and thinks he's American and who your mum has told you not to talk to'. In the US, however, Pontiac are known for a whole range of cars, some of which were even quite good. Unfortunately, Pontiac were also part of GM, who finally realised they had far too many car companies all making basically the same cars and needed to get their house in order or go under, end result of which was that Pontiac were closed down in 2010. Michael Knight would be worried about future parts supply for KITT, if only he existed, which, according to the titles of his own show, he doesn't.

P IS FOR PORSCHE

German sports-car company, officially called Dr. Ing. h.c. F. Porsche AG, but let's not get bogged down in that now. Founded by Ferdinand Porsche in 1931 as an engineering consultancy that won contracts to design tanks and what became the VW Beetle, and did some things in the era of roughly 1939–45 that they don't really mention much anymore. Anyway, after that, Porsche knuckled down to becoming a sports-car company, starting with the 356, which was launched in 1948. This was followed by the 911 of 1963, and Porsche carried on using and evolving this design even after the rest of the world had pretty much abandoned rear-engined cars because, as Richard Hammond once said, where a normal person who kept stubbing their toe on the bedside table would move the table, a Porsche engineer would keep redesigning their foot. Over the years they've tried mid-engined cars, they've tried front-engined cars, in recent times they've got into SUVs and saloons, and they're all very good. In their life to date they've also had a hand in the design and engineering of everything from fork-lift trucks and tractors to F1 cars and airliners, and these were probably very good too. But the sheer, bloody-minded perseverance with the 911 means it's now tremendously good indeed.

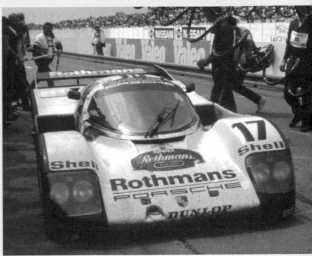

If you think range-extender hybrids are a new thing, you're wrong, because Ferdinand Porsche designed one in 1899.

Q IS FOR ...

QUARTER LIGHT

QUARTER

Q-CAR

QUAD

QUAD BIKE

QUALI

QOROS

Q7

QUATTRO

Q IS FOR ...

QUAD BIKE

Four-wheeled motorcycle used to round up animals and injure celebrities. Quad bikes are famed for their wayward nature, largely because many don't have a proper differential, which means they turn in a peculiar way. Riding a quad bike in the countryside makes you look like a farmhand, whereas riding one in a town makes you look like you're up to no good.

Q-CAR

A high-performance car that looks to the untrained eye like an ordinary saloon or hatchback and gives no clue as to how quick it is. The Mercedes 300SEL 6.3, for example, or the original Audi S8. The four-door Ford Sierra Sapphire Cosworth might have been considered a Q-car in the way the three-door Ford Sierra Cosworth absolutely wasn't. The Q-car name comes from Q-ships, ordinary-looking British merchant vessels fitted with concealed armaments, created to lure in and then attack German U-boats during the First World War. The Q in their name comes from the Irish port of Queenstown, from which many of them sailed.

QUALI

Irksome abbreviation for 'qualifying' much used by smug Formula 1 commentators as a way of saying, 'Oh yeah, I'm here at the F1 and you're not, and I get to hang out with the teams' engineers and pick up their slang, so there.' See also Box and Deg.

Q7

Large Audi SUV and highly effective magnet for literally the worst people in society.

QOROS

Chinese–Israeli joint-venture car company, founded in 2007 and based in Shanghai, which hired lots of Euro car designers and engineers in an attempt to create a Chinese-made car that could match Western design, dynamic and safety standards. Haven't set the world on fire so far, probably because their name sounds like a type of cheap 1980s aftershave.

QUARTER LIGHT

Small, triangular bit of glass at the leading or trailing edge of a car's side window and which in olden times could be opened to permit the ingress of fresh air and the egress of cigarette smoke, neither of which are fashionable anymore.

Q IS FOR QUATTRO

Word used by Audi, derived from the Italian for 'four', originally as a model name on their very first four-wheel-drive car and subsequently as a badge to show that variants of their other cars were fitted with four driven wheels. Not to be confused with Quatro, which is a fizzy, sticky reference that only people who were at school in the UK in the 1980s will get.

IS FOR

ROUNDABOUT

RENAULT

RHEOSTAT

REAR-ENGINED

ROAD MAP

RALLYING

RELIANT SCIMITAR

RED DIESEL

RAIN-SENSING WIPERS

R IS FOR RELIANT

Deceased British car maker, founded in 1935 in the automotive powerhouse of Tamworth. Specialised in funny little three-wheelers, leading up to the infamous Robin of 1973, staple of 1980s comedy routines and what lazy TV shows wheel out today when they have a car-related guest on and think they're being 'funny'. Some refer to this car as the 'Robin Reliant', thereby marking themselves as people who don't know what they're talking about, unless they have a weird brain block and also talk about 'Focus Fords' and 'iPhone Apples'. Reliant soldiered on for years with a bizarre and incompatible range that at one end had funny little low-powered three-wheelers (and, for a while, related four-wheelers like the Kitten and the Fox pick-up) and at the other had the rakish Scimitar sports car and, later, the not-quite-as-rakish-but-actually-quite-good SS1 roadster. Thanks to years of making cars from the stuff, Reliant were also pre-eminent in moulding fibreglass, and had a sideline in manufacturing bits for trains and vans and in the mid-1980s were contracted by Ford to build the fibreglass-bodied RS200 rally car. Unfortunately, by the time the 21st century came around everyone had generally agreed that the best number of wheels for a car was four and demand for Reliant's bedrock of funny little three-wheelers had waned. In 2001 the company stopped making cars for ever and its factory was demolished to make way for a housing estate. One of the roads is called Robin Close.

KITTEN

ROBIN

FOX

TANDY

In 1969 Reliant bought another company called Bond and later made the Bond Bug, which we used for our water-speed record car on *The Grand Tour*. I realise I probably should have mentioned this in the B section, but I'm afraid I forgot.

R IS FOR ...

RACING HARNESSES

Hardcore safety belts in three-, four-, five- and six-point styles, the main purposes of which are to keep the driver restrained in the event of an accident while also, in the case of male drivers, forcing the testes as far as possible into the body cavity.

RAIN-SENSING WIPERS

Vehicular technology in which the windscreen wipers detect rain or other moisture on the glass and activate automatically, as and when required. There are two possible answers to the question: How do rain sensors work?
1. They operate on the principle of refraction using an infra-red light inside the glass and a sensor that detects how much of that light is bounced straight back, on the basis that clean glass will bounce all the light straight back whereas water droplets will refract the light off in different directions, thereby sending less light back to the sensor, which then informs the control module that the glass needs wiping.
2. They don't.

RALLYING

A-to-B-format motorsport taking place on a wide range of terrain, from tarmac to gravel to snow, each stage seemingly chosen for narrowness, proximity to large objects and terrifying drops, and giving drivers the chance to really kick the tail out on the car in an exceptionally cool way. Rallying in the sense of who can get from here to there as quickly as possible has been going since the late 19th century, but rallying as we know it didn't really kick in until the 1950s and then took off in the 1960s as the plucky BMC Minis took on all-comers including, weirdly, large Citroëns and Porsche 911s. At this point the template was set for the sport, and from then on at least 80 per cent of the drivers had to be Finnish and called things like Liki Hovalovalenenen and Wanki Pantinantonenen, while all the co-drivers had to have solid names like Bob Stott and all the fans had to wear bobble hats.

RAN WHEN PARKED

Somewhat bollocks responsibility-dodging claim made in classified adverts for old cars that basically means 'This car will not start because I haven't touched it for seven years, a great deal of moisture has entered the electrical system and a mouse has eaten some of the wires.' Total nonsense. Because, let's be honest, even the *Titanic* was 'running' when it 'parked'.

RASPING

A word that in an automotive context can only – ONLY – be used to describe the exhaust note of an Alfa Romeo.

RE-GAS, JUST NEEDS A

Bold self-diagnosis made by 80 per cent of used-car sellers to suggest how simple it would be to make the air-conditioning in their wares fully operational again, begging the question: Why didn't you have it done ages ago then, eh?

REAL TIME

Baffling slogan written on the side of four-wheel-drive versions of the 1987 Honda Civic Shuttle 4WD. Turns out it was Honda's way of describing its automatic, on-demand four-wheel-drive system, which detected when the front wheels were slipping and directed some power to the rear wheels through a viscous coupling. Not that any of this was obvious to onlookers. They just thought it was part of some mad 1980s Japanese habit of writing random words on stuff, like people in Tokyo wearing T-shirts that said, 'Hey Mr Bernard, fondle my reasons.'

REAR-ENGINED

A car in which the motor is mounted behind the rear axle, bringing advantages to packaging the power unit close to the driving wheels while permitting luggage space under the bonnet and in some cases allowing better crash performance since the structure can be crushable without

having to contain the battering ram of a cylinder block in the event of a head-on smash. On the downside, rear-engined cars developed a fearsome reputation for sudden and unwanted oversteer and, after once being commonplace, fell out of fashion. Having a slight comeback these days, what with the Smart range and its Renault Twingo sister being rear-engined. VW were going to make the Up rear-engined too, but it all started to sound a bit complicated and expensive, so they didn't. Really, if your main priority when buying a car is that the engine must be right at the back, then just get a Porsche 911.

REAR-WHEEL DRIVE

Vehicle propulsion system in which the aft wheels receive power, leaving the front pair to take care of steering. Result is a dynamically 'pure' driving experience, plus the prospect of power oversteer if you want it. Downsides are that RWD usually needs the engine to be mounted lengthways and demands a propshaft to take torque to the back wheels, both of which eat into interior space. But the car might be more fun to drive. BMW certainly thought so, and made a great virtue of their purely rear-drive roots until they did some research and discovered that drivers of their cheaper cars had no idea which wheels were driven and didn't seem to care. That's why the 1 and 2 Series will be front-wheel drive in future. Because it turns out that keen drivers care about rear-wheel drive but estate agents don't.

RED DIESEL

Lower-tax agricultural fuel used by UK farmers in tractors and other farm vehicles and which is literally dyed red so it can be identified in spot checks by the taxman,

who has made it illegal to use in any vehicle that goes on the road. Also known as 'cherry juice', which sounds delicious, but isn't. No, really, it isn't.

R IS FOR RENAULT

French car maker founded in 1899 by the Renault brothers, which, unlike most companies of that age, went straight into car making rather than taking a run-up by building washing machines or flower presses or trousers first. One of Renault's early successes was to make taxis for Paris, a number of which were later commandeered by the government to transport troops to the front at Marne during the First World War. Renault went on to make tractors and tanks as well as cars and vans, gradually transforming from an upmarket concern building expensive things into a company of the people with affordable cars like the 4CV, the Dauphine and of course the Renault 4, which for years provided transport to the working people of France and the geography teachers of England. People think Citroën were innovative in the mid-20th century but Renault weren't far behind, just without the weirdness, and among other things they furthered the cause of the spacious family hatchback with the Renault 16 of 1965 and were in the vanguard of what we came to call the supermini with the 1972 Renault 5. Then in the '80s they became wilfully boring, only to go a bit nutty again with things like the windscreen-less Sport Spider, that Clio with the V6 in the middle and the completely batty Avantime. Renault also has a proud motorsport heritage in rallying and Formula 1, so try to remember that before you start moaning that your Megane has developed another rattle.

R IS FOR ...

RELIANT SCIMITAR

See Princess Anne.

REV COUNTER

Dashboard dial used to indicate engine speed. Also, a good name for the vicar in a romantic novel.

RHEOSTAT

You know that little wheel that adjusts the instrument lighting on your dashboard and that no one really uses but which car makers insist on fitting anyway? Well, that's called a rheostat. Basically, it's a variable resistor that controls the current in an electrical circuit and...
[That's quite enough for now thanks, James – JC]

RILEY

Dormant British car company started as a bicycle maker in 1890, bought by Morris in 1938, then sucked into British Leyland in 1968 and promptly killed off.

When BMW bought Rover in 1994 they decided Riley was actually quite a good thing and formulated some plans to bring the company back, then had second thoughts and didn't bother. They still own the name to this day.

RIMAC

Croatian electric-car maker founded in 2009 by youthful entrepreneur Mate Rimac. Most famous for the sensational 1,200-horsepower Concept One supercar, although Rimac are actually pre-eminent in electric car tech and have done work for Aston Martin, Jaguar and Koenigsegg. In fact, Porsche were so impressed by their knowhow that in 2018 they bought into the company.
[Well, this entry seems to have finished, so now I suppose you're going to crash it off a mountain, Hammond – JC]
[Funny. Reeeeally funny – RH]

ROAD MAP

What people called sat-nav in the old days and which, unlike sat-nav, lived in the footwell of your car where it would get trodden on and torn by passengers before magically disappearing whenever you needed it.

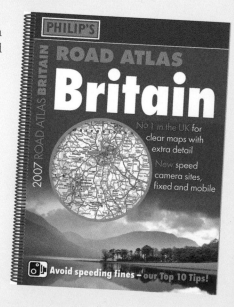

ROLLING ROAD

Rotating cylinder-based system that enables a car's performance and power output to be measured in static conditions and subsequently permits the owner of a highly tuned Skyline or Sierra Cosworth to quote an extremely high number while chanting the words 'at the wheels' and waving a piece of graph paper in your face until you hope he will just go away.

and effectively used on the Ford Cortina 1600E, although variations of the same design also appeared on the MGB, Rover P5B, Mini 1275GT and, weirdly, the original Range Rover. The thing about Rostyle wheels is that they were affordable, simple and really, really cool.

ROOF RACK

Metal structure attached to the top of a car to permit the carrying of extra loads and the subsequent depositing of these loads all over the A303. The old fashioned roof rack isn't very popular anymore on account of everyone having those smooth, enclosed roof boxes that do at least prevent your pants from becoming strewn across arterial roads just outside Yeovil.

> The last bit of this entry is wrong. My first car was a Cortina 1600E on Rostyle wheels and I think you'll find they were actually really, really, really, really, REALLY cool.

ROSTYLE WHEELS

Pressed-steel road wheel of the 1960s and '70s made by Rubery Owen & Co. (hence the name, which derives from 'Rubery Owen Style') to a basic design from the USA called the Magnum 500, as fitted to muscle cars like the Pontiac GTO. The Rostyle was most famously

ROUNDABOUT

Circular traffic-management feature called an 'island' in some parts of the UK and 'extremely confusing' in some parts of the USA.

RUST

Oxidisation of metal in contact with air and water, and the reason why you can now see the road from your 1970s Lancia while looking at the floor.

R IS FOR ROLLS-ROYCE

British car company that started selling cars in 1904 after the coming together of car salesman Charles Rolls and engineer Henry Royce (left). Soon developed a reputation for fine engineering and swiftly alighted on a slightly spooky naming policy with the Phantom and then the Silver Ghost, both traits that remain today. Became very successful, such that in 1931 they could afford to buy their rival, Bentley, and kill off their cars, so that for decades afterwards Bentleys were just Rolls-Royces with a slightly different grille. Rolls-Royce also became an engine maker for both planes and trains, but their cars remained as a standard bearer, much though they had to evolve with the times, introducing the smaller, cheaper

Silver Shadow in 1965 to supply a new generation of rock stars, showbiz impresarios and self-made millionaires who were drawn to Rollers and – shock! – didn't necessarily have chauffeurs. Went bust in 1971 after the aero business blew all the cash on developing the new-generation RB211 jet engine and the car division became independent, before being bought by Vickers and then, in 1998, by BMW in a hilarious grab, after VW bought the Rolls factory and Bentley but forgot to check that the rights to use the Rolls-Royce name were included, which they weren't. VW paid £430m and didn't get Rolls-Royce. BMW handed over just £40m and did. Quite a bargain, even if they then had to put their hand in their pocket for a from-scratch Phantom and a brand new, futuristic, half-buried factory in Goodwood, which is where the current Rolls-Royce range comes from, including that new SUV thing that looks like a taxi. Note that since Mr Rolls was a mere salesman and Mr Royce was the engineering maestro, some people don't like the nickname 'Rolls' and prefer to say their car is a 'Royce'. This is a good warning that you might want to make your excuses and find someone else to talk to.

Charles Rolls was the first Briton to be killed in an air crash involving powered flight when his Wright Flyer came down near Bournemouth in 1910.

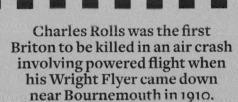

I can beat that, May. Around the time Henry Royce died in 1933 Rolls changed the background of their badge from red to black. People thought it was a mark of respect but actually it was because their customers said the red clashed with certain paint colours.

R IS FOR ROVER

Resting British car company, founded as a bicycle maker in 1878 before moving into motorcycles and then, in 1904, car manufacture. Bimbled around for many years before a pair of brothers called Wilks arrived at the company at the start of the 1930s and kicked it into shape, moving Rover upmarket of common-or-garden Fords and such like, becoming pre-eminent in gas turbines during the Second World War and then, once war was over, inventing the Land Rover. However, in the post-war years the main thing Rover was known for was stout, upright, high-quality motor cars, even if their ongoing and slightly bonkers project to fit gas turbines to road cars suggested the firm weren't as sensible as they seemed. The radical P6 of 1963 boasted some unusual engineering precisely because Rover always wanted it to have a gas turbine, although the most interesting engine it ever received was a lightweight V8, the design of which the Wilks brothers had wisely bought from Buick. In 1967 Rover got sucked into what became British Leyland, although it rode out this disappointing turn of events by coming up with the Range Rover and then the stylish SD1, although the wonky quality of both suggested that the leaden hands of BL had already infested their Solihull engineering block. Rover's reputation took a dive off the back of shoddy '70s reliability, enjoyed a resurgence in the '80s as Rover became one of the few surviving parts of the BL empire and felt the benefit of a partnership with Honda, then slumped again as BMW bought the company and everything got a bit National Trust as a Rover became the go-to car for people who probably wouldn't live to see its first MOT. In 2000 the Germans sold up to some Brummie businessmen who kept it hanging in there for five years, only to see the place collapse, with the wreckage sold to the Chinese. Don't weep, though, because the spirit of Rover was always in its original home of Solihull and that's where, to this day, Land Rovers and Range Rovers are made, and indeed where the Rover name quite literally lives, since BMW sold the rights to Ford who sold them to Jaguar Land Rover's overlords, Tata. They're probably not going to bring Rover back. Not unless they suddenly have some mad ideas about gas turbines.

S IS FOR ...

SAAB

SKODA

SUZUKI

SMART

SPORT MODE

SIXTEEN-VALVE

SELF-DRIVING CARS

STATION WAGON

SNOWMOBILE

S IS FOR SAAB

Defunct Swedish car company famed over the years for doing several interesting things, although making a profit and selling lots of cars were rarely among them. SAAB began in 1945 when Svenska Aeroplan AB (Swedish Aeroplane Ltd) decided to have a crack at cars and came up with a weird, teardrop-shaped prototype that evolved into the two-cylinder, two-stroke, front-wheel-drive SAAB 92 of 1949, so called because SAAB already had a numbering system for its planes and the last aircraft it had made was the 91. The odd little 92 set the template – unique, aeroplane-inspired design and a little bit of weirdness – for future SAABs, as the 92 evolved into the 93 and then the 95 and 96 (the 94 was a strange little sports car that never really, ahem, took off) and then into the all-new 99 of 1968. Although SAABs could seem odd, much of what the company did was driven by the precise requirements of their home market. So they were in the vanguard of front-wheel drive because it gave greater traction in slippery conditions, they were pioneers of safety because in Sweden you might skid off the road or clatter into an elk, they always fitted powerful heaters and auto-on heated seats to fight off harsh Scandinavian winters. They were also one of the first car companies to launch a turbocharged car, and even that seemed to come from a very Swedish desire to make more power without being so wasteful as to fit a larger engine. Perhaps because of this none-more-Scandinavian image, SAABs became the choice of intelligent, thoughtful people, and these customers stuck with them even when, in 1989, GM bought into the company and the cars became rehashed Vauxhalls with a Swedish accent. Unfortunately, SAAB kept ignoring GM's attempts to make them use as many generic bits as possible, and there simply aren't enough intelligent, thoughtful people in the world to make a business case for bespoke cars that sell in tiny numbers. That's why, in 2008, the Americans started looking for a way out and, after unsuccessfully trying to sell the company to Koenigsegg, in 2011 managed to offload it onto Spyker, who promptly got into financial trouble, saw a lifeline from a Chinese car company that didn't work out, collapsed, and then let the Chinese pick through the wreckage, a car-industry move known as 'doing a Rover'.

S IS FOR ...

SCANDINAVIA FLICK

Rallying trick in which the driver enters a corner by briefly flicking the steering in the wrong direction while lifting off the throttle and dabbing the brakes to unsettle the car, then steers in the right direction while re-applying power. Thus the car begins to drift neatly through the corner (if you're 19-time Finnish rally champion Petta Kakalakaboom) or begins to crash messily into a ditch (if you're not).

SCISSOR DOORS

Car doors that hinge upwards rather than outwards, most commonly seen on Lamborghinis and extensively modified Peugeot 306s owned by blokes called Daz. It's not clear why they're called scissor doors since they're rubbish at cutting paper, although you could probably take off a toenail if you were careless.

SELF-DRIVING CARS

Vehicles that can guide themselves to a destination without any human input. Hysterical newspaper reports and the growing number of cars that can crudely steer themselves might lead the casual observer to think that self-driving cars will be with us within a few months, or a year or two at most. They won't. They're being worked on, but so are jetpacks – and we've been promised those for ages too. There's a man who appears at public displays and flies one. He's followed at all times by his own personal fire engine, which tells you all you need to know about that. Frankly, self-driving cars are the same. It's a technology that's in its infancy and needs a great deal of refining before it can hope to negotiate downtown Manhattan or central London or the death road in Bolivia with the absolute and total reliability such systems need if they're to succeed. And another thing . . .
[Okay, James, all good points but you need to calm down and have a rest now, thanks – JC]

SEMI

What Americans call an articulated lorry. British people, stop giggling.

SENNA, AYRTON

Brazilian racing driver famed for being brilliant.

SENNA, MCLAREN

British sports car famed for being so brilliant they named it after a man who also was.

SICURSIV

See visrucis.

SIDE REPEATER

Small flashing light on the wing panel of a car (or on the actual door, if you're a first-generation Range Rover Sport or Mk1 Renault Laguna), which flashes in time with the front and rear indicators to provide greater visibility. At least, that's what it was until in 1998 Mercedes had the bright idea of moving it onto the door mirror and most other car makers decided to copy them, so now when some berk in a van smashes your mirror it's got even more gubbins inside it and costs three times as much to mend. Bah!

SILK UNDERPANTS

Luxurious sub-trouser item cited by Art Blakeslee, the chief designer of the first Renault Espace, as the swanko symbol of success all his team would be wearing once that car went on sale. Unfortunately, his confidence in the smash-hit potential of the Espace and its ability to enhance the undercracker situation among his staff was short-lived, as in its first month on sale the total number of Renault Espaces sold was nine.

SIXTEEN-VALVE

Thing that used to seem exciting in the 1980s because it usually denoted a four-cylinder engine with four valves per cylinder, which sounded terribly high-tech (even though the Triumph Dolomite Sprint had offered this in 1973) and ignored the fact that V8 engines have technically had 'sixteen valves' for years. Something car makers don't really mention anymore, because most four-cylinder engines have four valves per cylinder and it would sound as silly as bragging that your new model had disc brakes or reclining seats.

SKID PAN

Artificially low-friction area of tarmac used for car-control training, which, if archive footage from the 1970s is to be believed, was almost constantly occupied by an out-of-control double decker bus.

S IS FOR SEAT

S ociedad Española de Automóviles de Turismo (Spanish Society of Touring Cars). Iberian car maker founded in 1950 by the Spanish state and reliant for many years on building re-hashed Fiats. Then they had a row with Fiat and came up with their own car, which looked suspiciously like a Fiat Strada but was not, oh dear me no, señor, not likely. After that, they invented the first Ibiza, which definitely didn't look like a Fiat and was quite nice because they asked Giugiaro to design the outside and Porsche to help with the innards, and then in 1986 they got bought by VW, which has spent the last 30-odd years trying to make people believe that SEAT is a sort of Spanish Alfa Romeo, even though in general their cars just feel like ordinary VWs with a slightly spicier sauce.

> SEAT didn't just make old Fiats. In the '70s they also started making their own version of the Lancia Beta coupe. Then they fell out with the Italians and stopped again.

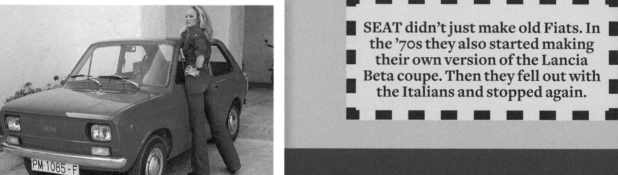

S IS FOR SKODA

Czech car maker founded in 1895 by Václav Laurin, who was an engineer, and Václav Klement, who wasn't. In fact, he was a bookseller. The early days in their workshop must have been a riot. 'Oi, Václav, pass me that spanner.' 'Not now, Václav, I'm reading.' Like a lot of Victorian-era car makers, Laurin & Klement actually started out making bicycles before moving into things with four wheels and an engine, and then getting bought by arms manufacturer Pizen Skodovka in 1925 to become Skoda. After the Second World War the company fell under the communistical curse of state control, peeking out from behind the Iron Curtain to sell a series of funny little rear-engined cars to the West, notable for their Renault-derived engine, which had an advanced aluminium block but a hefty cast-iron head, where other car makers did it the other way round, and the fact that, unlike most other Eastern Bloc cars of the era, they weren't without charm. Skoda came roaring into the modern age with the Favorit of 1987, which had its engine at the front, and then they started getting sucked into the VW Group in 1991, after which Skodas increasingly became re-engineered Volkswagens, but somehow more pleasant and likeable, a tradition they continue to this day.

S IS FOR ...

SLAM PANEL

Technical name for the cross beam that goes across the front of a car and usually carries the latch for the bonnet. Also a really good name for some kind of wrestling committee.

SMART

German maker of little cars, rooted in a plan by the people behind Swatch watches to build a stylish, customisable city runabout. Got Volkswagen on board in 1991, only for them to scarper two years later, at which point Mercedes stepped in and the whole thing was named Smart as a contraction of 'Swatch Mercedes Art', although Swatch ceased to be involved shortly after their first car came out in 1998. Have carried on making a funny little two-seater ever since, adding a roadster and a larger hatchback, then deleting them again and developing an SUV-style hatchback and not even bothering to put that on sale because what kind of idiot small-car maker would suddenly announce a massive hatchback. Apart from Mini. The main thing about the two-seater Smart, now in its third generation and named the fortwo, although everyone always calls it a 'Smart car', is that many people who own one believe it's short enough to be parked nose-in to the kerb in between other cars, even though it's patently obvious to anyone with a tape measure or indeed a set of eyes that it's not and is surely about to have its arse clobbered off by a bin lorry.

SNOWMOBILE

Motorbike with skids, let down by its inability to work anywhere but on snow. Used chiefly by Arctic types, James Bond and the racing driver Kimi Räikkönen, who once entered a snowmobile race in his native Finland when he was supposed to be preparing for the 2007 Formula 1 season and, presumably to avoid detection, gave his name as 'James Hunt'.

SPOILER

An aerodynamic device designed to disrupt or 'spoil' airflow, for example, directing airflow away from the sharp drop-off at the back of a car. Not to be confused with a wing, which is a thing that creates downforce, or some idiot on the internet accidentally giving away the plot twist in a film you've yet to watch.

SPORT MODE

Feature of many modern cars that in many cases sharpens throttle response, shortens gearshift time, stiffens the suspension, and makes the entire car feel both more sporty and more unpleasant.

SPORT UTILITY VEHICLE

Modern term for what used to be called a '4x4' or a 'Jeep', and which now describes any high-riding car that is not actually sporty at all, nor, in some cases, utilitarian, nor, frankly, much of a vehicle.

STATION WAGON

American word for 'estate car', derived from early 20th-century workhorses used to transport people and their luggage from the railway station to their final destination. These days they'd just use an SUV. For the entire journey.

STINGER

Scissor-hinged extendable strip of spikes that police people can throw out to ambush a fleeing car and burst its tyres, thereby bringing it to a stop or causing it to drive on the metal rims, making lots of sparks in a way that looks dramatic on those Cop! Stop! Now!-type TV programmes. Also a type of Kia that, because it's rear-wheel drive and has a limited slip diff, is also able to suddenly appear sideways across a road.

STOP

Simple instruction written on a red-framed octagonal sign all over the world, although not in the Canadian province of Quebec, where their devotion to being literally the most French people in the world means their stop signs say 'Arrêt', in contrast to actual France, where they say 'Stop' because France signed up to the treaty formulated at the Vienna Convention on Road Traffic in 1968, which aimed to standardise road signs as much as possible. So there.

STOPPING DISTANCE

Length of road needed for a car to stop from a given speed, as set out in the British Highway Code. However, The Highway Code says the stopping distance from 60 mph is 73 metres, but a McLaren Senna can stop from 60 in just 29 metres. That's not even half as much. Sounds like the Highway Code people need to do a special version for McLaren Senna drivers. Or at least stop basing their distance measurements on readings taken with a 1973 Vauxhall Viva on worn tyres.

S IS FOR SSANGYONG

Korean manufacturer founded in 1963 by the merger of Ha Dong-hwan Motor Workshop and Dongbang Motor Co. to form the Ha Dong-hwan Motor Company, before changing their name to Dong-A Motor. Honestly, please try to stop giggling. Dong-A Motor became part of the SsangYong industrial conglomerate and changed their name to . . . well, you know. SsangYong became famed for making a series of funny-looking 4x4s using some Mercedes bits and then really upped their game on the styling front by launching a range of extraordinarily unattractive vehicles, apparently styled by someone who had left their reading glasses in a cab. Bought by Daewoo in 1997, only to be sold again three years later, they went bust in 2009 before being rescued by Indian company Mahindra & Mahindra in 2011, so they could return to selling not-very-good-looking things for moderate prices. Little wonder they can now claim to be Korea's fourth-largest car maker.

S IS FOR **SUBARU**

Japanese car maker founded by Fuji Heavy Industries and named after the Japanese word for the Pleiades star cluster in the constellation of Taurus, six stars of which appear on the Subaru badge. Launched their first car in 1954 and have been at it ever since, selling things with weird features like boxer engines, standard four-wheel drive and frameless door windows in saloons, as if they didn't really know or care what the rest of the world was up to. Turns out that at least two of these things make for really good rally cars and all of these things make for often pleasant road cars, either with big spoilers and driven by yobs, or no spoilers and several Labradors in the back and driven by quiet, thoughtful people and country types. Subaru don't sell many cars in Europe but they're absolutely massive in the US, perhaps because their adverts there have dogs in them. Everyone likes adverts with dogs in them.

The Subaru Outback of a few years ago was one of only three cars in history that Hammond, May and I all like.

S IS FOR ...

STRAIGHT SIX

Engine layout in which six cylinders are arranged in a straight line. You probably guessed that from the name. Famed for their smoothness, straight sixes appeared in lots of cars, usually arranged lengthways and driving the rear wheels, although British Leyland, Daewoo and Volvo were all nutty enough to squeeze them crossways in front-wheel-drive models, then they fell out of favour because they tend to be very long and hard to accommodate. Only BMW didn't swerve from the path of straight-six loveliness, but the good news is that now Merc have come back to them and Jaguar will soon too. So that's nice.

SUNSHINE ROOF

Old British expression for a sunroof. Something car dealers used to make a big deal out of, even though it was just a small pop-up glass panel crudely stuck into a hole in the roof and, besides, there's hardly any sun in Britain anyway.

SUPER UNLEADED

Higher-octane fuel, required by some high-performance cars. Also what you give your normal car as a 'treat'.

SUPERCHARGER

Engine-driven compressor that forces more air into an engine to the benefit of output. Superchargers have some benefits over exhaust-driven turbochargers, notably their 'always-on' ability to start working without waiting for gases to build up and their amusing whirring noises, but they also make engines quite thirsty. Jaguar once claimed that on a 400bhp supercharged engine the supercharger at full chat took 70 horsepower to run. They should have taken it off, then they'd have had a 470bhp engine. No, wait, hang on . . .

SUPERLEGGERA

Literally, 'super light', a car-construction technique created by the Italian coachbuilder Touring based around lots of very thin steel tubes covered in an aluminium skin. Used by several glorious sports cars of the 1950s and '60s, including the Aston Martin DB4 and Lamborghini 350GT, actual Superleggera construction fell out of fashion but the name has been used by Lamborghini on a hardcore version of the Gallardo and more recently on the new Aston DBS, even though it weighs 1,845 kilos and is therefore not exactly 'light', even if it is really rather 'super'.

SYMMETRICAL FOUR-WHEEL DRIVE

Odd name given by Subaru to its all-wheel-drive system, derived from the fact that all the drive shafts are the same length, which they say is better for handling and not because the same amount of drive goes to all four wheels all the time so that they always turn at the same speed. That would be silly, and the car wouldn't be able to go around corners.

S IS FOR SUZUKI

Japanese company founded in 1909 as a loom maker. Had a brief stab at cars, got derailed by the Second World War and finally paid attention to the 'become car maker' bit of its to-do list in 1955 with something called the Suzulight, which had a 360cc engine and advanced features like front-wheel drive, double-wishbone suspension, and rack and pinion steering, paving the way for Suzuki to become a maker of revvy, clever little cars like the current Ignis and Swift, as well as funny, bouncy little off-roaders like the SJ and Jimny. A strangely likeable car company.

T

IS FOR ...

TALBOT

TACHOGRAPH

TRIUMPH

Turbo

TVR

TESLA

TOYOTA

T IS FOR ...

TACHOGRAPH

Device wired into a lorry that records speed, distance and time to keep a record of the driver's working hours. Hence the requirement for lorry drivers to take a 'tacho break' of at least 45 minutes in every four and a half hours, during which they may eat, sleep or purchase a pornographical magazine from one of those weird truck-stop places on the A1 that feels like there might be a dead body slumped behind the back wall.

TAXI

Motor vehicle and driver combination offering, in return for money, to take you to your destination or, in the case of British minicabs, somewhere quite near it. Also a word shouted by people trying to hail taxis in busy places like London and New York, but only in films.

TATA

Indian conglomerate and owner of Jaguar Land Rover. Also has its own car-making division that produced the Tata Indica, re-badged in Britain as the CityRover and quite possibly the worst car ever sold.

TAX DISC

Obsolete disc of thin paper displayed in the windscreen to show that your road tax, or Vehicle Excise Duty, as it's officially called, had been paid. The physical tax disc was abolished in 2014, thereby removing the requirement for car owners to tear around the tiny perforations to remove the disc from its square surroundings without ripping it, also known as the most casually perilous thing a grown human could do.

THINGS HANGING FROM THE REAR-VIEW MIRROR

A quick and reliable indicator that someone isn't very good at driving.

THREE-LITRE CAR

Shorthand for a car that can cover 100 kilometres on just three litres of fuel, the equivalent of 94 miles per gallon. Legend has it that many years ago Volkswagen got wind that Renault was working on a '3-litre' Clio and rushed forward their own plans to make a Lupo that could match this incredible fuel economy. Imagine their surprise when they discovered that Renault's plan had nothing to do with fuel economy and everything to do with building a bonkers mid-engined Clio with a 3-litre V6 in the middle.

THUNDERSLEY INVACAR

Small, plastic three-wheeler commissioned by the British government and leased to people with disabilities. The classic light-blue Invacar was a regular sight in the UK from 1972 until 2003, when the authorities took back all remaining examples and crushed them, although production actually ended in 1977 after concerns about their safety and some harsh words about their handling from, of all people, racing driver Graham Hill, who said they possessed 'the most unstable configuration of wheels you could have'.

Some Invacars were built by AC, the people who also made the Cobra. I'm sure there was a lot of crossover.

TOP GEAR

[Can't think of a definition for this one. Hammond? – JC]
[I've got nothing, sorry. James? – RH]
[Stumped. Soz. – JM]

TORQUES

Flippant, television-friendly way to refer to the torque output of an engine – that is to say, its twist action or how much rotational force it generates – generally expressed in the UK using the pound-foot measurement, where one pound-foot is the torque created by a one-pound weight acting one foot away from a pivot point. However, a more modern way to express torque is with the Newton metre (Nm), and you will often see this now used in car statistics from manufacturers. Interestingly … [That's quite enough of that, thanks, James – JC]

TOUTES DIRECTIONS

Literally, 'all directions'. Road sign invented by the French for when they can't be bothered to go into detail and just want you to get out of their town.

TRACTOR

Agricultural vehicle designed to travel extremely slowly across fields while pulling farm implements like ploughs and muck spreaders, and to travel barely any less slowly on roads while getting in the way of other things like cars and vans. Also one of the things Richard Hammond thinks about to achieve excitement.
[No it isn't – RH]
[Yeeees it is – JC]

TRANSAXLE

An integrated transmission and differential unit. Transaxle is often used in reference to a front-engined, rear-drive car in which the gearbox is mounted at the back for the benefit of weight distribution – e.g. Aston Martin DB11, Porsche 928, Corvette, etc. – or a front-engined, rear-drive car in which the gearbox is mounted at the back to allow the gear linkage to break and fall onto the propshaft with catastrophic results, including the release of the loudest and most terrible noise ever experienced on earth – e.g. an Alfa Romeo GTV6 owned by Jeremy Clarkson in the 1980s.

TRANSIT

Ford commercial vehicle first launched in 1965, whose name has become a generic word for 'van' in the UK. The actual Transit was almost called the V-series, which wouldn't have worked half as well and may have led online parcel trackers to tell you 'your item is in V-series', causing you to wait in all day feeling tense and edgy.

T IS FOR TALBOT

Defunct car maker founded in Britain in 1903 and originally called Clément-Talbot after its founders, Adolphe Clément-Bayard and Major Charles Chetwynd-Talbot, Earl of Shrewsbury, Earl of Waterford and later Viscount Ingestre. Lovely lad. His dad was a plumber. Enjoyed great success in racing, got bought and split, thereby entering a bizarre Lotus-in-F1 situation in which there were two Talbots for a while, died a death, got sold to Simca, who got sold to Chrysler Europe, who got bought by Peugeot, who brought the name back as

a way to re-badge the Chrysler range it had inherited when it got the keys in 1978 and which led to the all-new (but really a Peugeot underneath) Talbot Tagora of 1981, which remains the least-curvy car ever made, as well as one of the least successful. Talbot stopped making cars in 1987, although the name soldiered on attached to vans until 1994, when it ceased altogether. What would Major Charles Chetwynd-Talbot, Earl of Shrewsbury, Earl of Waterford and later Viscount Ingestre say about that, eh? Nothing. He died in 1921.

T IS FOR TESLA

American electric car company founded in 2003 and selling its first car, the Lotus Elise-based Roadster, in 2008. Now offers the Model S executive car, the Model X SUV with those cool gullwing back doors, and the entry level Model 3. Run by Elon Musk, who is famously not fond of criticism and has a history of responding to any perceived slight by calling people rude names on Twitter or getting his lawyers involved, so let's just say all Teslas are absolutely excellent and leave it there.

T IS FOR TOYOTA

Japanese giant founded in 1933 as a car-making offshoot of the Toyoda Automatic Loom Works. Built their first prototypes in 1935 and promptly had them blessed in a Buddhist ceremony, which clearly worked because things have gone well for Toyota ever since, barring a little pause in car production for the Second World War, a brief period of near-bankruptcy thereafter and, more recently, that whole massive recall for sticking accelerator pedals a few years ago. But, in general, Toyota have done okay and often outsell VW to claim the title of world's biggest car maker. For years Toyotas have majored on being well made and reliable, thanks in part to the famed Toyota Production System, an integrated approach to every aspect of making a car that has been developed and refined over 70 years, with an emphasis on avoiding inconsistency and overburden while avoiding wastage. It's the benchmark for reliable, lean production, which is probably why when Porsche was at their lowest and least profitable ebb in the late '80s they called upon Toyota to help them make cars more efficiently. It's very easy to think of Toyota as a maker of high-quality, reliable but ultimately very boring cars, until you remember all the good stuff they've come up with. The 2000GT, the Supra, the Corolla AE86, the MR2, the Celica, the Lexus LFA, the GT86 . . . the list of interesting, sporty stuff is long, and that's before you get to some of the technically brilliant unsporty things like the original Lexus LS400 and the bonkers Mk1 Previa. These are the high spots, but in the meantime the boring, everyday saloons get half of America and Australia and Japan to work, and the indestructible pick-ups and vans keep most developing nations moving. Basically, without Toyota the entire world would grind to a halt.

T IS FOR ...

TRUNK

What Americans call a 'boot', unless you are an American elephant, in which case it means 'nose'.

TUNNEL

Underground road passing through a mountain or under a body of water, the main purpose of which is to allow you to drop your window, knock it down a gear and give a huge blast of exhaust sound to annoy James May.

TURBO

Exhaust-driven compressor feeding more air to the engine to give more power. Also a word that used to be slapped on cars so equipped because it sounded exciting and cool. In fact, back in the 1980s almost everything seemed to have the word 'turbo' on it. Turbo sunglasses. Turbo aftershave. Turbo hygienic hand wipes. There was nothing turbo on a product couldn't do. Except use exhaust gases to drive an impeller to force more air and therefore fuel into the combustion chamber to increase output at the potential expense of driveability due to the lag sometimes experienced between applying more throttle and sufficient exhaust gases being released to drive the turbocharger. That only applied to cars.

TURD SPEED

The amount by which a motorist is exceeding the speed limit because they urgently need a poo.

T IS FOR TRIUMPH

Defunct-ish British company founded as a bicycle importer in 1885 before becoming a motorcycle maker and then, in 1921, a car company. Evolved into a famous maker of sports cars, notably the TR roadsters, which started with the TR2 of 1953, but it was in saloons that Triumph was ahead of its time, cramming a straight six into the Herald to create a quick, compact, rear-drive Vitesse in a move that would later become BMW's stock-in-trade, previewing the idea of a smooth executive car with the elegantly long-wheelbase 2000 and 2500, and refining the idea of a lively compact sports saloon with the Dolomite Sprint. Unfortunately, Triumph was bought by Leyland in 1960 and sucked into the subsequent mire of BL, which, along with industrial problems in the 1970s and some comical cock-ups like the undercooked V8 engine in their gloriously stylish Stag, sealed Triumph's fate. Their last sports car, the TR7, was killed off in 1981 and in the same year the Triumph name was used to rebadge a Honda saloon as the Acclaim. When that car died in 1984, so did Triumph. Which is a shame, because they did some good things and could have been a British rival to a certain sporty German saloon-car maker. Ironically enough, the actual Triumph name is now owned by BMW.

T IS FOR TVR

British sports-car company established in 1946 by Trevor Wilkinson and gaining their title from a truncated version of the founder's first name. So it's a good job he wasn't called Trint. TVR found their feet turning out funny little fibreglass machines with other people's engines in them and developed the sort of brassy, no-nonsense approach you'd expect from a company based in Blackpool, culminating in increasingly unruly performance from the cars and naked women on their British motor-show stand. The company hit their peak under moustachioed Yorkshireman Peter Wheeler, who bought the place in 1981, installed Rover V8s in the wedge-shaped range he'd inherited and then caused a thousand 1990s' bankers to soil themselves with his boisterous range of stunning sports cars, including the beautiful Griffith, the almost unspellable Chimaera and the elegant Cerbera, each packed with performance and design detailing bonkers enough to distract you from all the bits that fell off or didn't work. Unfortunately, in 2004 Wheeler sold up to Russian businessboy Nikolai Smolensky, everything went to cock and the TVR factory closed in 2006. Happily, in 2013 a proper grown up called Les Edgar bought the TVR name from the bored bizchild and is busily getting ready to bring it back with an all-new Griffith. However, until he starts claiming that the door releases are brass nipples hidden in the headlights and that a dog designed the front bumper, he's not going to get close to the nuttiness of TVR in their heyday.

U

IS FOR ...

UNIPART
USED CAR
UNDERSTEER
UMBERSLADE INTERCHANGE
USED-CAR DEALER

UNLEADED PETROL

U IS FOR ...

UMBERSLADE INTERCHANGE

The place where the M40 and M42 motorways meet in the British West Midlands, notable because no matter which way you approach the merger heading south, the traffic joining from the other motorway always appears to be doing 320 mph.

UNDERSTEER

Handling characteristic in which the vehicle follows a wider line than should be delivered by the amount of steering input. Typically engineered into mainstream cars, as it's considered more 'safe' than oversteer because you can see the tree you're going to hit.

UNIPART

Car-parts company, formerly part of British Leyland. Used to boast the slogan, 'The answer's yes, now what's the question?' although they might change their tune if you asked, 'Does my Allegro work properly?' or 'Did you pay Angela Lansbury £200 to flay a rat with a bike chain?'

UNLEADED PETROL

What petrol is often referred to, acknowledging that it no longer has lead in it after everyone realised this was a stupid idea and was damaging important parts of people, such as their brains. Petrol doesn't naturally have lead in it, so 'unleaded' is misleading and makes as much sense as referring to black tea as 'unmilked'.

USED CAR

A vehicle that has already had at least one owner before you, something that the official used-car schemes of most car makers attempt to mask with phrases like 'pre-owned' or 'ready loved' and daft brand names like 'Approved selection' or 'Experiences'. No, it's just a car that someone else has farted and sneezed in.

USED-CAR DEALER

Badly regarded member of the community, typically operating out of a redundant petrol station or tatty former Rover dealership, and whose chief arsenal in the battle to win the public's attention is to wait for a nice day and then open all the hatchbacks.

V

IS FOR ...

VLOGGER

VAG

VENT SPEWS

VINTAGE CAR

VENTILATED SEATS

VAUXHALL

VISRUCIS

V8 VOLVO

VAN

VOLKSWAGEN

V IS FOR...

V8

An engine with eight cylinders arranged in two banks of four, typically at a 90-degree angle. Not as smooth as a straight six, nor as compact as a V6, nor as economical as a four, and yet somehow the greatest engine format in the world.

You'd think the V8 came from America, but in fact it was invented by the French.

VAG

Titter-inducing abbreviation for the official name of the Volkswagen Group, which controls Audi, Bentley, Bugatti, Lamborghini, Porsche, Skoda, SEAT and Volkswagen itself. Often wrongly claimed to stand for Volkswagen Audi Group, but it doesn't. It stands for Volkswagen Aktiengesellschaft, which basically means Volkswagen plc, but in German.

VAN

Commercial vehicle capable of moving large loads and, in the case of Mercedes Sprinters on British motorways, travelling at 278 mph.

VAN DRIVER'S THANK YOU

Double flash of the hazard lights to show gratitude to a motorist who has just permitted you to pull out or merge in. A van driver's thank you must never exceed two flashes, otherwise it just looks like you're having an actual breakdown.

VENT SPEWS

Little 'hairs' on new tyres, caused by tiny bits of molten rubber going up the ventilation pipes on the moulds. A better name would be 'pnuebs'.

VENTILATED SEATS

Comfort feature in which tiny fans and perforated upholstery permit a delightful cooling effect on the back and bottom in hot weather, while giving the chance to covertly blow flatulence over the people in the back seat.

VINTAGE CAR

A pre-Second World War car typified by a general appearance of something from one of those black and white films in which everything seems to drive at 190 mph with 19 policemen hanging off it, controls that are all in the wrong order and largely on the outside, and a complete resistance to the boom in classic car prices because no one can relate to them as totems of their childhood and the only people who find them interesting are increasingly dead.

VISRUCIS

See Sicursiv.

VLOGGER

In the car world, a vlogger is generally a well-spoken young man who pops up on YouTube, referring to the audience as 'guys' and posting a series of videos announcing that he's thinking of buying a very expensive supercar, debating what colour it should be, declaring that it's arrived, deciding to have it wrapped, and then telling everyone he's selling it. DON"T FORGET TO LIKE AND SUBSCRIBE, GUYS!

VOICE ACTIVATION

Car technology that is claimed to allow the driver to adjust the radio, alter the temperature, input sat-nav destinations and make telephone calls, all with just the push of a button and the natural spoken word, but which in reality forces even the most clearly spoken person to keep shouting, 'Set destin … no, set destination … SET destination … Manches … destination SET … Man-chester … CANCEL … set DESTINATION … M … A … N … no, cancel … CANCEL … Set. Destin. Ation … WHY ARE YOU CALLING MIKE FLETCHER?'

V IS FOR VAUXHALL

British company founded in 1857 as a maker of marine engines and named after the Vauxhall area of London, where their factory was situated. Went into car making in 1903 and were famed in their earliest days for the sportiness of their designs. Bought by General Motors in 1925 and turned into a less glamorous seller of everyday transport, a mildly drab tradition they continue to this day. For many years Ford and Vauxhall have fought it out to be Britain's favourite everyday car maker, but somehow Ford has always had something more. Sporty Fords are revered, whereas to say you have a sporty Vauxhall just sounds funny, even though many of them in the past, like the HP Firenza, Manta GTE and Lotus Carlton, were actually good fun. Some of their ordinary cars, like the Mk2 Cavalier, kicked Ford's bottom too, but still and to this day Vauxhall sounds like the kind of thing nobody dreams of, apart from Peugeot who dreamt of it so much they bought the whole company.

V IS FOR VOLKSWAGEN

German car maker founded in 1937, although they generally prefer to talk about their history post-1945 when a British army officer called Major Ivan Hirst got their bomb-damaged factory running again and offered its contents, including a funny little bug-like car, to a British car industry consortium led by Lord Rootes, who snorted in derision and said the funny little bug-like car would be 'quite unattractive to the average motorcar buyer'. The car was the Beetle and they sold 21 million of them. In fact, the success of the strange bug-like car was partly down to its unattractiveness, and VW's American ad agency used to play on its role as a sort of anti-car to great effect. For years Volkswagen was basically a one-car company (plus the Type 2 van and minibus), and when they tried to replace the Beetle they kept getting it wrong until, in 1974, they made one last desperate roll of the dice and came up with the Golf, which was a smash hit, kick-started the idea of a versatile, do-all Euro

hatch for the '80s, and gradually killed off the slow, noisy, archaic Beetle. From that modern foundation, VW went on to create brilliant, sturdy, attractive everyday cars like the Polo and Up, and excellent things in the sensibly sporty area such as the Corrado and of course the Golf GTI. Like Toyota, VW are one of the bedrocks of the car landscape and help to keep the world moving. Although let's not forget they also lie about emissions and gas monkeys, the horrible bastards.

V IS FOR VOLVO

Swedish company that came into existence in 1911, but only as a name for Sweden's leading brand of ball bearings. Morphed into a car maker, selling their first model in 1927, majoring then and for ever more on stout cars formulated for the roads of Sweden. Volvo became known for safety, not least because their adverts used to bang on about it when no other car maker appeared interested in such a thing, and also for its large estate cars, without which the British antiques trade of the 1970s and '80s would have ground to a halt. Or just sold very small things. Old Volvo 240s and 760s have bags of retro appeal now, which is funny because at the time they were seen as stodgy and boring, while Volvo themselves have slowly sexed up their image with smoother styling, some turbo-nutter engines and excellent japes like entering an estate car in the British Touring Car Championship. Now make a range of handsome, versatile models and vie with Skoda for the title of world's most casually pleasant car company.

W

IS FOR ...

WIPERS

WATCH THIS!

WRITE-OFF

WD-40

WHEEL SHUFFLER

WHEELBASE

WANKEL

WHITEWALL TYRES

W IS FOR...

WANKEL

Schoolboy-amusing design of engine invented by Felix Wankel in the 1920s and also known as a rotary because instead of reciprocating pistons it has one or more rotors that are driven round by internal combustion, leading to incredible smoothness and strong power from a very compact, small capacity unit. On the downside, these engines are rather thirsty and use a lot of oil, and also Felix Wankel was a massive Nazi.

Cars that have used Wankel engines include various Mazdas such as the RX-7 and RX-8, the NSU Ro 80 and a couple of very short-lived Citroëns. Plenty of choice for anyone who wants a car with an engine designed by a Nazi.

WATCH THIS!

The kind of exclamation that often precedes a car accident.

WD-40

Water displacer, lubricant, rust inhibitor and general-purpose spray-it-around-the-engine-in-the-hope-that-the-bastard-thing-starts-working. So-called because it was the 40th mixture that the Rocket Chemical Company in San Diego came up with in their attempt to create a water-displacement formula. Hats off to whoever greeted the despondent mood in the lab upon the failure of the 39th try with the words, 'No, hang on, everyone. Let's give it one more go . . .'

WHEELBASE

The distance between the front and rear wheels of a car, generally the same across all variants of a single model unless you are a Renault 21, which, bizarrely, came in four different wheelbases depending on engine size and whether it was a saloon or an estate.

WHEEL SHUFFLER

A member of any advanced-driver club, the key tenet of which is always to avoid crossing your arms when turning the steering wheel as if power steering hasn't been invented and you really want to look like a busy tit.

WHITEWALL TYRES

Strange automotive fashion from the USA, in which a black tyre has a strip of white rubber around its outer face. Very popular in the 1950s and '60s, now about as fashionable as other quirky things from the past, i.e. due a comeback among millennials any day now.

WIPERS

Arcing metal arms fitted with rubber blades, designed to clear water and muck off a car's windscreen. Generally come in twos, although Citroëns used to have one, as did the 1980s Jaguar XJ6 (because the chief engineer was a Citroën fan), and in the '70s MG got around the problem of the MGB failing to meet new American rules on the amount of screen the wipers had to wipe by fitting US models with three wipers. To sum up: wipers come in various numbers, and you're only in trouble if there's none.

WRITE-OFF

What your car often becomes shortly after you have said, 'Watch this!'

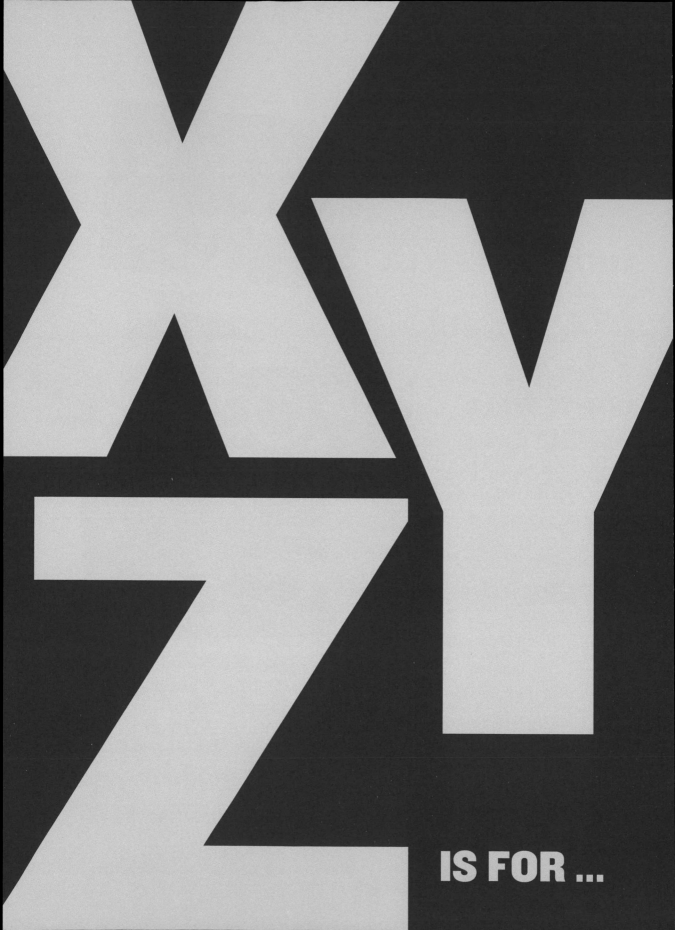

Z

YELLOW HEADLIGHTS

XEDOS

ZEBRA CROSSING

XENON

XX | Z-AXLE

X IS FOR ...

XEDOS

Short-lived subset of Mazdas intended to be more upmarket and luxurious. Swiftly deleted once it was discovered that no one much wanted more upmarket and luxurious Mazdas.

XX

Mysterious-sounding codename given to not one but two British car projects, the first being what turned into the Rover 800, the second a top-secret plan to restyle the Jaguar XJS, which then turned into the Aston Martin DB7. Also a band, presumably not named after the codenames of the Rover 800 or Aston DB7 project.

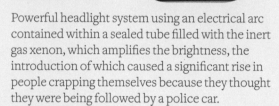

XENON

Powerful headlight system using an electrical arc contained within a sealed tube filled with the inert gas xenon, which amplifies the brightness, the introduction of which caused a significant rise in people crapping themselves because they thought they were being followed by a police car.

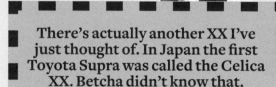

There's actually another XX I've just thought of. In Japan the first Toyota Supra was called the Celica XX. Betcha didn't know that.

Y IS FOR...

YELLOW HEADLIGHTS

French tradition dating back to 1936 when their government made yellow lights compulsory, superficially because l'Académie des sciences said they were less dazzling, but more likely because France feared invasion and thought yellow lights would make it easier to distinguish between invaders' vehicles and those of their own people (ignoring that it would also make the French easier to spot and shoot at). After the Second World War the French stuck with yellow lights, still maintaining the 'undazzling' line until finally, in 1993, EU rule makers persuaded them this was largely bollocks, which was a shame because, whatever the claimed benefits, the main thing is that for over 50 years the Champs-Élysées at night looked incredibly cool.

Z IS FOR ...

Z-AXLE

BMW name for a multi-link rear-suspension system first used on the nutty Z1 sports car of 1989 and thereafter seen on pretty much all their cars, plus, when adapted for front-wheel drive, the Rover 75 and new Mini. So-named not, as often claimed, because it's Z-shaped but because in German it's a *zentralpunktgeführte, sphärische Doppelquerlenkerachse* design, but of course you knew that already.

ZEBRA CROSSING

Road crossing so-named because of the dark and white markings on the road. Not to be confused with PED XING, which is how they alert you to road crossings in the USA and is also the name of the fourth-largest plastic-producing region in China.

ZIL

Russian vehicle maker most famous for its huge limousines, as used by high-ranking members of the Communist Party, hence the dedicated road spaces in Moscow that only senior government types could use were known as 'ZiL lanes'. ZiL stopped making the limousine in 2002 and ceased truck production 10 years later, but in 2010 the government ordered them to make three more limos for the Moscow Victory Day parade, as captured in a bleakly brilliant documentary called *The Last Limousine*. Now ZiL is no more, and most of their vast factory has been turned into a housing estate, rather echoing the fate of that other communist bastion, Longbridge in Birmingham.

Z IS FOR ZAGATO

Italian styling house, founded in 1919 and famed in the modern era for coming up with designs that don't necessarily conform to conventional ideas of attractiveness. Let's be honest, some of their stuff is horrible. And even the good ones are sometimes marred by mad details, like the Alfa Junior Z of 1969, which was a nice design spoilt by having needlessly massive rear-wheel arches, like someone who's had their hair cut too high around their ears. Mind you, Zagato is still going after 100 years, so maybe there's money in making cars that almost look right if you stand at a very specific angle and squint a bit.

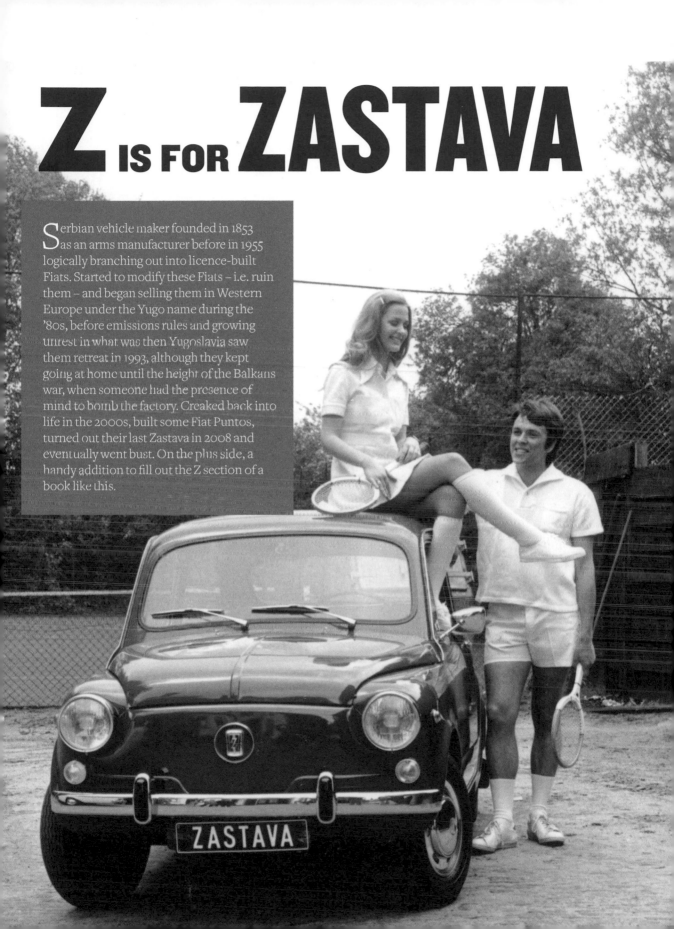

Z IS FOR ZASTAVA

Serbian vehicle maker founded in 1853 as an arms manufacturer before in 1955 logically branching out into licence-built Fiats. Started to modify these Fiats – i.e. ruin them – and began selling them in Western Europe under the Yugo name during the '80s, before emissions rules and growing unrest in what was then Yugoslavia saw them retreat in 1993, although they kept going at home until the height of the Balkans war, when someone had the presence of mind to bomb the factory. Creaked back into life in the 2000s, built some Fiat Puntos, turned out their last Zastava in 2008 and eventually went bust. On the plus side, a handy addition to fill out the Z section of a book like this.

HarperCollins*Publishers*
1 London Bridge Street
London SE1 9GF

www.harpercollins.co.uk

First published by HarperCollins*Publishers* 2018

10 9 8 7 6 5 4 3 2 1

A catalogue record of this book is available from the
British Library

ISBN 978-0-00-825788-0

Printed and bound at GPS Group

MIX
Paper from
responsible sources
FSC™ C007454

This book is produced from independently certified FSC™ paper
to ensure responsible forest management.

For more information visit: www.harpercollins.co.uk/green

All photos courtesy of the featured manufacturers/
companies/organisations via Newspress, with
thanks, and also the Giles Chapman Library.
Our additional thanks to Audi for images of
Procon-Ten and to Neal Anderson for pictures
of his Thundersley Invacar.

The following images © Shutterstock.com
p19 (mini), p23 (seat), p29 (clock), p29 (sign),
p38 (car sketch), p39 (sign), p39 (car port),
p42 (CB radio), p42 (car clock), p55 (Daimler),
p58 (car wash), p58 (pump), p64 (ambulance),
p73 (flange), p78 (fog lights), p83 (fuel light),
p83 (fuse), p96 (highway code), p100 (horn),
p107 (roundabout sign), p107 (Italy flag), p135
(diff lock), p135 (axles), p144 (texture), p167
(nipple), p190 (microphone), p202 (Henry
Royce), p205 (Rover), p215 (stop sign), p253
(zebra crossing); Paula Solloway/Alamy Stock
Photo p23 (cyclist); Sipa Press/Rex/Shutterstock
p61 (Delorean); © BBC Photo Library p235
(Frank Butcher).